# NOBEL PRIZE WINNERS IN PHYSICS

# NOBEL PRIZE WINNERS IN PHYSICS

**Arun Agarwal**

**A P H PUBLISHING CORPORATION**
**ANSARI ROAD, DARYA GANJ**
**NEW DELHI-110 002**

*Published by*
S.B. Nangia
**A P H Publishing Corporation**
4435-36/7, Ansari Road, Daryaganj
New Delhi 110002
Ph.: 23274050
E-mail : aphbooks@gmail.com

2023



Rs. 2995/-

*Printed at*
**Balaji Offset**
Navin Shahdara, Delhi 110032

*Dedicated To All Those*
*Who Have Enlightened Me*
*And Made*
*It Possible For Me To Achieve It*
*And Those Who*
*Wish To Carry This Light In Future*

# ACKNOWLEDGEMENT

At the foremost, I would like to thank the God, give me strength, determination and insight, needed for the successful completion of this book.

Words are inadequate in translating deep sentiment of gratitude to my respectable, sagacious, eminent and honourable ideal teacher cum supervisor Prof. K.G. Varshney, Ph.D. D.Sc., Ex-Chairman, Department of Applied Chemistry, Aligarh Muslim University, Aligarh (India), who always encourage me to move ahead. I am very fortunate to get an opportunity to work with such a dynamic personality. I hope his incessant guidance would support me to face the challenges during my whole life. My vocabulary is inadequate to express my emotions, regards and sincere thanks towards him.

I would like to extend my sincerest and heartiest indeptedness, that naturally sprouts from the core of my heart, to respected Mrs Nirmal Varshney. I especially wish to acknowledge her polite nature and fruitful suggestions which give the specific shape to this work.

I read a number of National and International Journals of last many years, for which I am highly obliged to the seminar libraries of Department of Physics, Chemistry, Zoology and Botany, Aligarh Muslim University, Aligarh (India).

I have approached many peoples and visited many National level scientific institutions and libraries for collecting the valuable information.

No words are adequate enough to express my warm heartful thanks to my parents, my sister Mrs. Krati Goyal, my brother-in-law Mr. Dev Suman Goyal, without whom this herculean task would have not completed, whose sacrifice and ceaseless affection, restored my confidence. When I faltered, prayers have always been a soul of inspiration and guidance during this work.

Last but not least I am profoundly grateful to publishers for their whole hearted co-operation in bringing out this book in record time.

**Arun Agarwal**

# Contents

## Part II



## Part III



## Part IV



## Part V

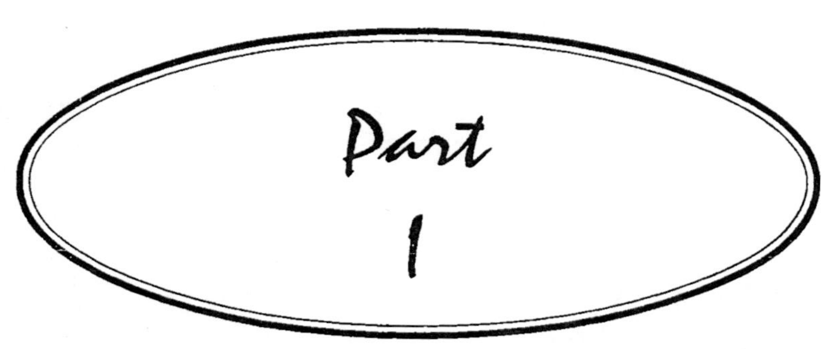
Part
1

- *Alfred Nobel–Man Behind the Prizes*
  *Nobel Stamps*
- *Alfred Nobel (Portrait)*
- *Alfred Bernhard Nobel (Biography)*
- *Alfred Nobel–His Life and Work*
- *Nobel Foundation*
- *Nobel Prizes*
- *The Nobel Medals and the Medal for the Memorial Prize in Economic Sciences*
- *The Nobel Diplomas*
- *Nobel Prize Amount in Swedish Crowns*
- *List of Nobel Prize Winners (Physics)*
  *(Year, Name, Nationality, Work)*
- *List of Nobel Prize Winners (Physics)*
  *(Year, Name, Nationality)*
- *Illustrations*

  *The obverse side of the Nobel Prize medals for Physics, Chemistry, Physiology or Medicine, and Literature.*

  *The reverse side of the Nobel Prize medal awarded for both Physics and Chemistry.*

  *The reverse side of the Nobel Prize medal for Physiology or Medicine.*

  *Virtual Tour of the Royal Swedish Academy of Sciences*

  *Nobel Stamps*

# Alfred Nobel—Man behind the Prizes

Alfred Nobel was born in 1833 in Stockholm, Sweden to a family of engineers. His family was descended from Olof Rudbeck, the best-known technical genius of Sweden's 17th century era as a great power in northern Europe. At age 9, he moved with his family to Russia where he and his brothers were given first class education in the humanities and natural sciences by private teachers.

**Nobel invented dynamite in 1866 and later built up companies and laboratories in more than 20 countries all over the world. A holder of more than 350 patents, he also wrote poetry and drama and even seriously considered becoming a writer.**

The idea of giving away his fortune was no passing fancy for Nobel. Efforts to promote peace were close to his heart and he derived intellectual pleasure from literature, while science built the foundation for his own activities as a technological researcher and inventor.

On November 27, 1895, Nobel signed his final will and testament at the Swedish-Norwegian Club in Paris. He died of a cerebral haemorrhage in his home in San Remo, Italy on December 10, 1896.

**NOBEL STAMPS**

Alfred Nobel-1946

Alfred Nobel-1997

Alfred Nobel

*"After his own experiments led him to the lucrative invention of dynamite, Alfred Nobel established a fund to reward other innovators "contributing most materially to the benefit of mankind." The Nobel Prizes are among the most highly regarded awards an individual can receive. They are given in the areas of chemistry, physics, physiology or medicine, literature, and peace, and in 1968, an award in economics was introduced in Nobel's honour. The awards reflect the interdisciplinary interests of the man himself; in addition to performing valuable chemical research, Nobel spoke several languages, travelled widely, and wrote poetry."*

# ALFRED BERNHARD NOBEL

*b. Oct. 21, 1833, Stockholm, Sweden*
*d. Dec. 10, 1896, San Remo, Italy*

Swedish chemist, engineer, and industrialist who *invented dynamite* and other, more powerful explosives and who also founded the Nobel Prizes.

Alfred Bernhard Nobel was the fourth son *of Immanuel and Caroline Nobel.* Immanuel was an inventor and engineer who had married Caroline Andrietta Ahlsell in 1827. The couple had eight children, of whom only Alfred and three brothers reached adulthood. Alfred was prone to illness as a child, but he enjoyed a close relationship with his mother and displayed a lively intellectual curiosity from an early age. He was interested in explosives, and he learned the fundamentals of engineering from his father. Immanuel, meanwhile, had failed at various business ventures until moving in 1837 to St. Petersburg in Russia, where he prospered as a manufacturer of explosive mines and machine tools. The Nobel family left Stockholm in 1842 to join the father in St. Petersburg. Alfred's newly prosperous parents were now able to send him to private tutors, and he proved to be an eager pupil. He was a competent chemist by age 16 and was fluent in English, French, German, and Russian, as well as Swedish.

Alfred Nobel left Russia in 1850 to spend a year in Paris studying chemistry and then spent four years in the United States working under the direction of John Ericsson, the builder of the ironclad warship Monitor. Upon his return to St. Petersburg, Nobel worked in his father's factory, which made military equipment during the Crimean War. After the war ended in 1856, the company had difficulty switching to the peacetime production of steamboat machinery, and it went bankrupt in 1859.

Alfred and his parents returned to Sweden, while his brothers Robert and Ludvig stayed behind in Russia to salvage what was left of the family business. Alfred soon began experimenting with explosives in a small laboratory on his father's estate. At the time, the only dependable explosive for use in mines was black powder, a form of gunpowder. A recently discovered liquid compound, *nitroglycerin* (Pure nitroglycerin is a colourless, oily, somewhat toxic liquid having a sweet, burning taste. *It was first prepared in 1846 by the Italian chemist Ascanio Sobrero* by adding glycerol to a mixture of concentrated nitric and sulfuric acids.

The hazards involved in preparing large quantities of nitroglycerin have been greatly reduced by widespread adoption of continuous nitration processes. Nitroglycerin, with the molecular formula $C_3H_5(ONO_2)_3$, has a high nitrogen content 18.5%, was a much more powerful explosive, but it was so volatile that it could not be handled with any degree of safety.

Nevertheless, Nobel in 1862 built a small factory to manufacture nitroglycerin, and at the same time he undertook research in the hope of finding a safe way to control the explosive's detonation. In 1863 he invented a practical detonator consisting of a wooden plug inserted into a larger charge of nitroglycerin held in a metal container; the explosion of the plug's small charge of black powder serves to detonate the much more powerful charge of liquid nitroglycerin. This detonator marked the beginning of Nobel's reputation as an inventor as well as the fortune he was to acquire as a maker of explosives. In 1865 Nobel invented an improved detonator called a blasting cap; it consisted of a small metal cap containing a charge of mercury fulminate that can be exploded by either shock or moderate heat. The invention of the blasting cap inaugurated the modern use of high explosives.

Nitroglycerin itself, however, remained difficult to transport and extremely dangerous to handle. So dangerous, in fact, that Nobel's nitroglycerin factory blew up in 1864, killing his younger brother Emil and several other people. Undaunted by this tragic accident, Nobel built several factories to manufacture nitroglycerin for use in concert with his blasting caps. These factories were as safe as the knowledge of the time allowed, but accidental explosions still occasionally occurred. Nobel's second important invention was that of dynamite in 1867. By chance, he discovered that nitroglycerin was absorbed to dryness by kieselguhr, a porous siliceous earth, and the resulting mixture was much safer to use and easier to handle than nitroglycerin alone. Nobel named the new product dynamite ( from Greek dynamis, "power") and was granted patents for it in Great Britain (1867) and the United States (1868). Dynamite established Nobel's fame worldwide and was soon put to use in blasting tunnels, cutting canals, and building railways and roads.

Nobel, Alfred Bernhard: Alfred Nobel, portrait by Emil Osterman, 1915; in the Nobel Foundation, Stockholm.

In the 1870s and '80s Nobel built a network of factories throughout Europe to manufacture dynamite, and he formed a web of corporations to produce and market his explosives. He also continued to experiment in search of better ones, and in 1875 he invented a more powerful form of dynamite, blasting gelatin, which he patented the following year. Again by chance, he had discovered that mixing a solution of nitroglycerin with a fluffy substance known as nitrocellulose results in a tough, plastic material that has a high water resistance and greater blasting power than ordinary dynamites. In 1887 Nobel introduced *ballistite,* one of the first nitroglycerin smokeless powders and a precursor of cordite. Although Nobel held the patents to dynamite and his other explosives, he was in constant conflict with competitors who stole his processes, a fact that forced him into protracted patent litigation on several occasions.

Nobel's brothers Ludvig and Robert, in the meantime, had developed newly discovered oilfields near Baku (now in Azerbaijan) along the Caspian Sea and had themselves become immensely wealthy. Alfred's worldwide interests in explosives, along with his own holdings in his brothers' companies in Russia, brought him a large fortune. In 1893 he became interested in Sweden's arms industry, and the following year he bought an ironworks at Bofors, near Varmland, that became the nucleus of the well-known Bofors arms factory. Besides explosives, Nobel made many other inventions, such as artificial silk and leather, and altogether he registered more than 350 patents in various countries.

Nobel's complex personality puzzled his contemporaries. Although his business interests required him to travel almost constantly, he remained a lonely recluse who was prone to fits of depression. He led a retired and simple life and was a man of ascetic habits, yet he could be a courteous dinner host, a good listener, and a man of incisive wit. He never married, and apparently preferred the joys of inventing to those of romantic attachment. He had an abiding interest in literature and wrote plays, novels, and poems, almost all of which remained unpublished. He had amazing energy and found it difficult to relax after intense bouts of work. Among his contemporaries, he had the reputation of a liberal or even a socialist, but he actually distrusted democracy, opposed suffrage for women, and maintained an attitude of benign paternalism toward his many employees. Though Nobel was essentially a pacifist and hoped that the destructive powers of his inventions would help bring an end to war, his view of mankind and nations was pessimistic.

By 1895 Nobel had developed angina pectoris, and he died of a *cerebral hemorrhage* at his villa in San Remo, Italy, in 1896. At his death his worldwide business empire consisted of more than 90 factories manufacturing explosives and ammunition. The opening of his will, which he had drawn up in Paris on Nov. 27, 1895, and had deposited in a bank in Stockholm, contained a great surprise for his family, friends, and the general public. He had always been generous in humanitarian and scientific philanthropies, and he left the bulk of his fortune in trust to establish what came to be the most highly regarded of international awards, the Nobel Prizes.

We can only speculate about the reasons for Nobel's establishment of the prizes that bear his name. He was reticent about himself, and he confided in no one about his decision in the months preceding his death. The most plausible assumption is that a bizarre incident in 1888 may have triggered the train of reflection that culminated in his bequest for the Nobel Prizes. That year Alfred's brother Ludvig had died while staying in Cannes, France. The French newspapers reported Ludvig's death but confused him with Alfred, and one paper sported the headline ***"Le marchand de la mort est mort"*** ("The merchant of death is dead.") Perhaps Alfred Nobel established the prizes to avoid precisely the sort of posthumous reputation suggested by this premature obituary. It is certain that the actual awards he instituted reflect his lifelong interest in the fields of physics, chemistry, physiology, and literature. There is also abundant evidence that his friendship with the prominent Austrian pacifist Bertha von Suttner inspired him to establish the prize for peace.

Nobel himself, however, remains a figure of paradoxes and contradictions: a brilliant, lonely man, part pessimist and part idealist, who invented the powerful explosives used in modern warfare but also established the world's most prestigious prizes for intellectual services rendered to humanity.

# Alfred Nobel - His Life and Work

Alfred Nobel was born in Stockholm on October 21, 1833. His father immanuel Nobel was an engineer and inventor who built bridges and buildings in Stockholm. In connection with his construction work Immanuel Nobel also experimented with different techniques for blasting rocks.

Alfred's mother, born Andriette Ahlsell, came from a wealthy family. Due to misfortunes in his construction work caused by the loss of some barges of building material, Immanuel Nobel was forced into bankruptcy the same year Alfred Nobel was born. In 1837 Immanuel Nobel left Stockholm and his family to start a new career in Finland and in Russia. To support the family, Andriette Nobel started a grocery store which provided a modest income. Meanwhile immanuel Nobel was successful in his new enterprise

**Andriette Nobel** **Immanuel Nobel**

in St. Petersburg, Russia. He started a mechanical workshop which provided equipment for the Russian army and he also convinced the Tsar and his generals that naval mines could be used to block enemy naval ships from threatening the city.

The naval mines designed by Immanuel Nobel were simple devices consisting of submerged wooden casks filled with gunpowder. Anchored below the surface of the Gulf of Finland, they effectively deterred the British Royal Navy from moving into firing range of St. Petersburg during the Crimean war (1853–1856). Immanuel Nobel was also a pioneer in arms manufacture and in designing steam engines.

Successful in his industrial and business ventures, Immanuel Nobel was able, in 1842, to bring his family to St. Petersburg. There, his sons were given a first class education by private teachers. The training included natural sciences, languages and literature. By the age of 17 Alfred Nobel was fluent in Swedish, Russian, French, English and German. His primary interests were in English literature and poetry as well as in chemistry and physics. Alfred's father, who wanted his sons to join his enterprise as engineers, disliked Alfred's interest in poetry and found his son rather introverted. In order to widen Alfred's horizons his father sent him abroad for further training in chemical engineering. During a two year period Alfred Nobel visited Sweden, Germany, France and the United States. In Paris, the city he came to like best, he worked in the private laboratory of Professor T. J. Pelouze, a famous chemist. There he met the young Italian chemist Ascanio Sobrero who, three years earlier, had invented nitroglycerine, a highly explosive liquid. Nitroglycerine was produced by mixing glycerine with sulfuric and nitric acid. It was considered too dangerous to be of any practical use. Although its explosive power greatly exceeded that of gunpowder, the liquid would explode in a very unpredictable manner if subjected to heat and pressure. Alfred Nobel became very interested in nitroglycerine and how it could be put to practical use in construction work. He also realized that the safety problems had to be solved and a method had to be developed for the controlled detonation of nitroglycerine. In the United States he visited John Ericsson, the Swedish-American engineer who had developed the screw propeller for ships. In 1852 Alfred Nobel was asked to come back and work in the family enterprise which was booming because of its deliveries to the Russian army. Together with his father he performed experiments to develop nitroglycerine as a commercially and technically useful explosive. As the war ended and conditions changed, Immanuel Nobel was again forced into bankruptcy. Immanuel and two of his sons, Alfred and Emil, left St. Petersburg together and returned to Stockholm. His other two sons, Robert and Ludvig, remained in St. Petersburg. With some difficulties they managed to salvage the family enterprise and then went on to develop the oil industry in the southern part of the Russian empire. They were very successful and became some of the wealthiest persons of their time.

After his return to Sweden in 1863, Alfred Nobel concentrated on developing nitroglycerine as an explosive. Several explosions, including one (1864) in which his brother Emil and several other persons were killed, convinced the authorities that nitroglycerine production was exceedingly dangerous. They forbade further experimentation with nitroglycerine within the Stockholm city limits and Alfred Nobel had to move his experimentation to a barge anchored on Lake Malaren. Alfred was not discouraged and in 1864 he was able to start mass production of nitroglycerine. To make the handling of nitroglycerine safer Alfred Nobel experimented with different additives. He soon found that mixing nitroglycerine with silica would turn the liquid into a paste which could be shaped into rods of a size and form suitable for insertion into drilling holes. In 1867 he patented this material under the name of dynamite. To be able to detonate the dynamite rods he also invented a detonator (blasting cap) which could be ignited by lighting a fuse. These inventions were made at the same time as the diamond drilling crown and the pneumatic drill came into general use. Together these inventions drastically reduced the cost of blasting rock, drilling tunnels, building canals and many other forms of construction work.

The market for dynamite and detonating caps grew very rapidly and Alfred Nobel also proved himself to be a very skilful entrepreneur and businessman. By 1865 his factory in Krummel near Hamburg, Germany, was exporting nitroglycerine explosives to other countries in Europe, America and Australia. Over the years he founded factories and laboratories in some 90 different places in more than 20 countries. Although he lived in Paris much of his life he was constantly travelling. Victor Hugo at one time described him as "Europe's richest vagabond". When he was not travelling or engaging in business activities Nobel himself worked intensively in his various laboratories, first in Stockholm and later in Hamburg (Germany), Ardeer (Scotland), Paris and Sevran (France), Karlskoga (Sweden) and San Remo (Italy). He focused on the development of explosives technology as well as other chemical inventions, including such materials as synthetic rubber and leather, artificial silk, etc. By the time of his death in 1896 he had 355 patents.

Intensive work and travel did not leave much time for a private life. At the age of 43 he was feeling like an old man. At this time he advertised in a newspaper "Wealthy, highly-educated elderly gentleman seeks lady of mature age, versed in languages, as secretary and supervisor of household." The most qualified applicant turned out to be an Austrian woman, Countess Bertha Kinsky. After working a very short time for Nobel she decided to return to Austria to marry Count Arthur von Suttner. In spite of this Alfred Nobel and Bertha von Suttner remained friends and kept writing letters to each other for decades. Over the years Bertha von Suttner became increasingly critical of the arms race. She wrote a famous book, *Lay Down Your Arms* and became a prominent figure in the peace movement. No doubt this influenced Alfred Nobel when he wrote his final will which was to include a Prize for persons or organizations who promoted peace. Several years after the death of Alfred Nobel, the Norwegian Storting (Parliament) decided to award the 1905 Nobel Peace Prize to Bertha von Suttner.

**Bertha von Suttner**

Alfred Nobel's greatness lay in his ability to combine the penetrating mind of the scientist and inventor with the forward-looking dynamism of the industrialist. Nobel was very interested in social and peace-related issues and held what were considered radical views in his era. He had a great interest in literature and wrote his own poetry and dramatic works. The Nobel Prizes became an extension and a fulfilment of his lifetime interests.

Many of the companies founded by Nobel have developed into industrial enterprises that still play a prominent role in the world economy, for example Imperial Chemical Industries (ICI), Great Britain; Societe Centrale de Dynamite, France; and Dyno industries in Norway. Toward the end of his life, he acquired the company AB Bofors in Karlskoga, where Bjorkborn Manor became his Swedish home. Alfred Nobel died in San Remo, Italy, on December 10, 1896. When his will was opened it came as a surprise that his fortune was to be used for Prizes in Physics, Chemistry, Physiology or Medicine, Literature and Peace. The executors of his will were two young engineers, Ragnar Sohlman and Rudolf Lilliequist. They set about forming the Nobel Foundation as an organization to take care of the financial assets left by Nobel for this purpose and to coordinate the work of the Prize-Awarding Institutions. This was not without its difficulties since the will was contested by relatives and questioned by authorities in various countries.

## The Nobel Foundation

A private institution established in 1900 based on the will of Alfred Nobel. The Foundation manages the assets made available through the will for the awarding of the Nobel Prize in Physics, Chemistry, Physiology or Medicine, Literature and Peace. It represents the Nobel institutions externally and administers informational activities and arrangements surrounding the presentation of the Nobel Prize. The Foundation also administers Nobel symposia in the different prize areas.

# NOBEL PRIZES

**I. Introduction:** *Nobel Prizes,* annual monetary awards granted to individuals or institutions for outstanding contributions in the fields of physics, chemistry, physiology or medicine, literature, international peace, and economic sciences. The Nobel prizes are internationally recognized as the most prestigious awards in each of these fields. The prizes were established by Swedish inventor and industrialist *Alfred Bernhard Nobel,* who set up a fund for them in his will. The first Nobel prizes were awarded on *December 10, 1901,* the fifth anniversary of Nobel's death.

In his will, Nobel directed that most of his fortune be invested to form a fund, the interest of which was to be distributed annually "in the form of prizes to those who, during the preceding year, shall have conferred the greatest benefit on mankind." He stipulated that the interest be divided into five equal parts, each to be awarded to the person who made the most important contribution in one of five different fields. In addition to the three scientific awards and the literature award, a prize would go to the person who had done "the most or the best work for fraternity among nations, for the abolition or reduction of standing armies, and for the holding and promotion of peace congresses." Nobel also specified certain institutions that would select the prize winners. The will indicated that "no consideration whatever shall be given to the nationality of the candidates, but that the most worthy shall receive the prize."

*In 1968 the **Riksbank,** the central bank of Sweden, created an **economics prize** to commemorate the bank's 300th anniversary.* This prize, called the *Nobel Memorial Prize* in Economic Science, was *first awarded in 1969.* The bank provides a cash award equal to the other Nobel prizes.

**II. Nobel Foundation** : In 1900 the Nobel Foundation was established to manage the fund and to administer the activities of the institutions charged with selecting winners. The fund is controlled by a board of directors, which serves for two-year periods, and consists of six members:– five elected by the trustees of the awarding bodies mentioned in the will, and the sixth appointed by the Swedish government. *All six members are either Swedish or Norwegian citizens.*

In his will, Nobel stated that the prizes for *physics and chemistry* would be *awarded by the Swedish Academy of Sciences,* the prize for *Physiology or Medicine by the Karolinska Institute in Stockholm,* the *Literature prize by the Swedish Academy in Stockholm,* and the *Peace prize by a five-person committee elected by the Norwegian Storting (Parliament).* After the economics prize was created in 1968, the Swedish Academy of Sciences has held the responsibility of selecting the winners of that award.

All the prize-awarding bodies have set up Nobel committees consisting of three to five people who make recommendations in the selection process. Additional specialists with expertise in relevant fields assist the committees. The Nobel committees examine nominations and make recommendations to the prize-awarding institutions. After deliberating various opinions and recommendations, the prize-awarding bodies vote on the final selection, and then they announce the winner. The deliberations and voting are secret, and prize decisions cannot be appealed.

**III. Prizes :** A prize for achievement in a particular field may be awarded to an individual, divided equally between two people, or awarded jointly among two or three people. According to the Nobel Foundation's statutes, the prize cannot be divided among more than three people, but it can go to an institution. A prize may go unawarded if no candidate is chosen for the year under consideration, but each of the prizes must be awarded at least once every five years. If the Nobel Foundation does not award a prize in a given year, the prize money remains in the trust. Likewise, if a prize is declined or not accepted before a specified date, the Nobel Foundation retains the prize money in its trust.

The prize amounts are based on the annual yield of the fund capital. In 1948 Nobel prizes were about $32,000 each; in 1997 they were about $1 million each. In addition to a cash award, each prizewinner also receives a *gold medal* and a *diploma* bearing the winner's name and field of achievement. Prizewinners are known as *Nobel laureates.*

**IV. Selection of Prizewinners :** Nominations of candidates for the prizes can be made only by those who have received invitations to do so. In the fall of the year preceding the award, Nobel committees distribute invitations to members of the prize- awarding bodies, to previous Nobel prize winners, and to professors in relevant fields at certain colleges and universities. In addition, candidates for the prize in literature may be proposed by invited members of various literary academies, institutions, and societies. Upon invitation, members of governments or certain international organizations may nominate candidates for the peace prize. The Nobel Foundation's statutes do not allow individuals to nominate themselves. Invitations to nominate candidates and the nominations themselves are both confidential.

Nominations of candidates are due on February 1 of the award year. Then, Nobel committee members and consultants meet several times to evaluate the qualifications of the nominees. The various committees cast their final votes in October and immediately notify the laureates that they have won.

**V. Prize Ceremonies:** The prizes are presented annually at ceremonies in *Stockholm, Sweden, and in Oslo, Norway, on December 10,* the anniversary of Nobel's death. In Stockholm, the *king of Sweden* presents the awards in physics, chemistry, physiology or medicine, literature, and economic sciences. The *peace prize* ceremony takes place at the University of Oslo in the presence of the *king of Norway.* After the ceremonies, Nobel Prize winners give a lecture on a subject connected with their prize- winning work. The winner of the peace prize lectures in Oslo, the others in Stockholm. The lectures are later printed in the Nobel Foundation's annual publication, ***Les Prix Nobel*** (The Nobel Prizes).

# The Nobel Medals and the Medal for the Memorial Prize in Economic Sciences

—Birgitta Lemmel

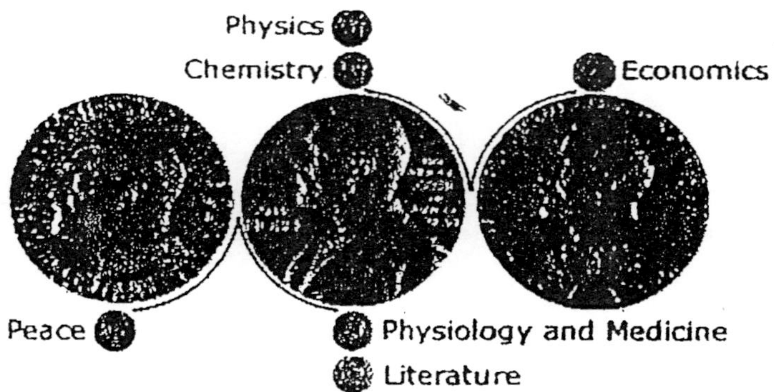


According to the Statutes of the Nobel Foundation, given by the King in Council on June 29, 1900, "the prize-awarding bodies shall present to each prize-winner an assignment for the amount of the prize, a diploma, and a gold medal bearing the image of the testator and an appropriate inscription."

The medals for Physics, Chemistry, Physiology or Medicine and Literature were modelled by the Swedish sculptor and engraver Erik Lindberg and the Peace medal by the Norwegian sculptor Gustav Vigeland. The medal for the Sveriges Riksbank (Bank of Sweden) Prize in Economic Sciences in Memory of Alfred Nobel (established in 1968 in connection with the 300th anniversary of the Bank of Sweden), was designed by Gunvor Svensson-Lundqvist.

The front side of the three "Swedish" medals (Physics and Chemistry, Physiology or Medicine, and Literature) is the same, featuring a portrait of Alfred Nobel and the years of his birth and death in Latin - NAT-MDCCC XXXIII OB-MDCCC XCVI. Alfred Nobel's face on the Peace medal and on the medal for the Economics Prize has different designs. The main inscription on the reverse side of all three "Swedish" Nobel Prize medals is the same: "Inventas vitam juvat excoluisse per artes," while the images vary according to the symbols of the respective prize-awarding institutions. The Peace medal has the inscription "Pro pace et fraternitate gentium" and the Economics medal has no quotation at all on the reverse.

Up to 1980 the "Swedish" medals, each weighing approximately 200 g and with a diameter of 66 mm, were made of 23-karat gold. Since then they have been made of 18-karat green gold plated with 24-karat gold.

Today the "Swedish" medals are cast by Myntverket - the Swedish Mint - in Eskilstuna and the Peace medal by Den Kongelige Mynt - the Royal Mint - in Kongsberg, Norway.

The Nobel medals have had the same design since 1902. Why not since 1901, when the first Prizes were awarded? In early 1901 the young and talented Swedish sculptor and engraver Erik Lindberg - later Professor Erik Lindberg - had been entrusted with the task of creating the three "Swedish" Nobel medals, while the Norwegian medal - the Peace medal - had been entrusted to the Norwegian sculptor Gustav Vigeland. The designs of the reverse sides of the "Swedish" Nobel medals were not finalized in time for the first Award Ceremony in 1901. We gather from Erik Lindberg's correspondence with his father Professor Adolf Lindberg that each of the 1901 Laureates received a "temporary" medal - a medal bearing the portrait of Alfred Nobel, cast in a baser metal - as a memento until the "real" medals were finished. The first of these medals was not completed and cast until September 1902.

During the years 1901-02 Erik Lindberg was living in Paris. He was influenced by modern French medal engravers of that period, such as the masters Roty, Chaplain, Tasset and Vernon. The portrait on the front of the Swedish medals was completed in time. It was reduced in October 1901 at Janvier's in Paris and the final punching took place in Stockholm. The reason for the delay was that the symbols on the reverse of the medals had to be approved by each Prize-Awarding institution, which was not without controversy. After lengthy discussions by letter, Erik Lindberg decided to return to Stockholm in November 1901 in order to present his ideas in person. His proposals were then all accepted, and he was finally able to produce the plaster casts for the reverse sides, which were then reduced for the final metal-stamping dies.

As Gustav Vigeland was a sculptor and not a medal engraver, Erik Lindberg was asked to make the dies for the Peace medal. His reductions were based on Vigeland's designs.

On all "Swedish" Nobel medals the name of the Laureate is engraved fully visible on a plate on the reverse, whereas the name of the Peace Laureate as well as that of the Winner for the Economics Prize is engraved on the edge of the medal, which is less obvious. For the 1975 Economics Prize winners, the Russian Leonid Kantorovich and the American Tjalling Koopmans, this created problems. Their medals were mixed up in Stockholm, and after the Nobel Week the Prize Winners went back to their respective countries with the wrong medals. As this happened during the Cold War, it took four years of diplomatic efforts to have the medals exchanged to their rightful owners.

On December 10 at the Prize Award Ceremony in Stockholm, His Majesty the King hands each Laureate a diploma and a medal. The Peace Prize, i.e. diploma and medal, is presented on the same day in Oslo by the Chairman of the Norwegian Nobel Committee in the presence of the King of Norway. The Irish poet William Butler Yeates wrote the following in "The Bounty of Sweden" (The Cuala Press, Dublin, 1925) after receiving the Nobel Prize for Literature in 1923:

"All is over, and I am able to examine my medal, its charming, decorative, academic design, French in manner, a work of the nineties. It shows a young man listening to a Muse, who stands young and beautiful with a great lyre in her hand, and I think as I examine it, I was good-looking once like that young man, but my unpractised verse was full of infirmity, my Muse old as it were; and now I am old and rheumatic, and nothing to look at, but my Muse is young'."

**Literature Laureate in 1986**
**Wole Soyinka's medal box**

There are many rumours of what happened to the Nobel medals of three Nobel Laureates in Physics during World War II: the medals of the Germans Max von Laue (1914) and James Franck (1925), and of the Dane Niels Bohr (1922). Professor Bohr's Institute of Theoretical Physics in Copenhagen had been a refuge for German Jewish physicists since 1933. Max von Laue and James Franck had deposited their medals there to keep them from being confiscated by the German authorities. After the occupation of Denmark in April 1940, the medals were Bohr's first concern, according to the Hungarian chemist George de Hevesy (also of Jewish origin and a 1943 Nobel Laureate in Chemistry), who worked at the institute. In Hitler's Germany it was almost a capital offense to send gold out of the country. Since the names of the Laureates were engraved on the medals, their discovery by the invading forces would have had very serious consequences. To quote George de Hevesy (Adventures in Radioisotope Research, Vol. I, p. 27, Pergamon, New York, 1962), who talks about von Laue's medal: "I suggested that we should bury the medal, but Bohr did not like this idea as the medal might be unearthed. I decided to dissolve it. While the invading forces marched in the streets of Copenhagen, I was busy dissolving Laue's and also James Franck's medals. After the war, the gold was recovered and the Nobel Foundation generously presented Laue and Frank with new Nobel medals." de Hevesy wrote to von Laue after the war that the task of dissolving the medals had not been easy, as gold is "exceedingly unreactive and difficult to dissolve." The Nazis occupied Bohr's institute and searched it very carefully but they did not find anything. The medals quietly waited out the war in a solution of aqua regia. de Hevesy did not mention Niels Bohr's own Nobel medal but documents in the Niels Bohr Archive in Copenhagen show that Niels Bohr's Nobel medal, as well as the Nobel medal of the 1920 Danish Laureate in Physiology or Medicine, August Krogh, had already been donated to an auction held on March 12, 1940 for the benefit of the Fund for Finnish Relief (Finlandshjälpen). The medals were bought by an anonymous buyer and donated to the Danish Historical Museum in Fredriksborg, where they are still kept. Regarding the Nobel medals of von Laue and Franck, the Niels Bohr Archive has a letter from Niels Bohr dated January 24, 1950, about the delivery of gold to the Royal Swedish Academy of Sciences in Stockholm relating to these two medals. The Nobel medals had been kept in the chemical substance in such a way that the Royal Mint in Stockholm preferred to strike new medals instead of trying to get them out of their wrapping. The proceedings of the Nobel Foundation on February 28, 1952, mention that Professor Franck received his recoined medal at a ceremony at the University of Chicago on January 31, 1952.

# The Nobel Diplomas

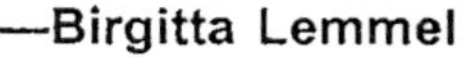

**—Birgitta Lemmel**

**Richard E. Smalley, Nobel Laureate in Chemistry 1996**

First published May 27, 1999: The festival day of the Nobel Foundation is on the 10th of December, the anniversary of the death of the testator. The Prize Award Ceremony for the Nobel Prizes in Physics, Chemistry, Physiology or Medicine and Literature as well as for the Bank of Sweden Prize in Economic Sciences in Memory of Alfred Nobel - takes place at the Stockholm Concert Hall. At this event, His Majesty the King of Sweden, hands each Laureate a diploma, a medal and a document confirming the Prize amount, which in 1999 will total SEK 7.9 million (about USD 1 million) per full Prize. The Nobel Peace Prize is presented on the same day at the Oslo City Hall by the Chairman of the Norwegian Nobel Committee in the presence of the King of Norway.

The prize-awarding bodies decide the design of the diplomas. The Royal Swedish Academy of Sciences is responsible for the Physics and Chemistry diplomas, and since 1969 also for the Economic Sciences diploma. The Nobel Assembly at Karolinska institutet is responsible for the Physiology or Medicine diploma, the Swedish Academy for the Literature diploma and the Norwegian Nobel Committee for the diploma presented to the winners of the Peace Prize. Nowadays, the "Swedish" diplomas have a uniform binding, provided by the bookbindery Falth & Hassler (earlier Hässlers Bokbinderi). This was not the case initially, since the various prize committees decided the artistic design of the diplomas based on their own wishes and resources. The Refsum bookbinding firm was responsible for binding the "Norwegian" diplomas until 1986, when the bookbinding firm of Kjell-Roger Josefson took over.

The artistic design of the diplomas has varied over the years, but the text has always followed the same pattern in the Swedish and Norwegian languages, respectively. The "Swedish" diplomas largely carry the same text, stating the person or persons to whom the prize-awarding body has decided to present the year's Prize plus a citation explaining why. The Norwegian diploma, on the other hand, has never included a Prize citation.

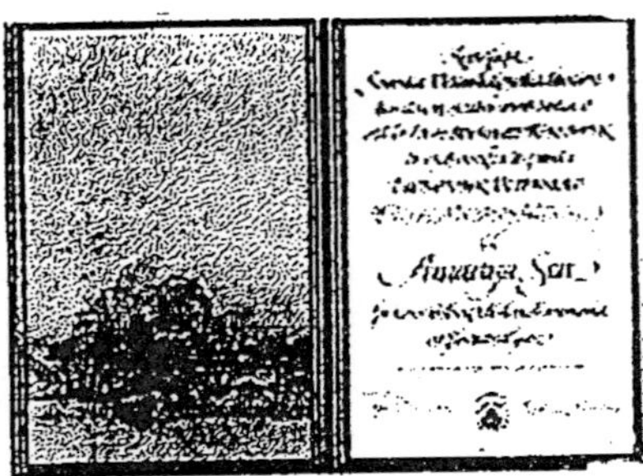

**Diploma of Amartya Sen, Winner of the 1998 Economics Prize Artist - Bengt Landin; Calligrapher**

**—Annika Rücker**

The Royal Swedish Academy of Sciences diplomas have been created by many artists since 1901: Sofia Gisberg (1901-1926); Ella Waldenstrom, Karin Ageman, Elsa Örtengren-Norêen, Björn Landström and Bertil Kumlien (1927–70); Gunnar Brusewitz (1970–73); Karl-Axel Pehrson (1974); Tage Hedqvist (1975-76); Sven Ljungberg (1977–89); Philip von Schantz (1990–93); and Bengt Landin (1994–98); and Nils G. Stenqvist from 1999. Very often, these diplomas are characterized by an annual theme - birds, flowers, vases etc.–rather than an individual design referring to the Laureates. In 1969, Reinhold Ljunggren created the diplomas for the First Economic Sciences Prize winners, Ragnar Frisch and Jan Tinbergen. Since 1970 the Physics, Chemistry and Economic Sciences diplomas have been designed by the same artist for each year.

**A Physiology/ Medicine diploma, the general look since 1965**

During the years 1901–64, the Physiology or Medicine diplomas were decorated with art works. In the first six years, 1901–06, these were created by artist and architect Agi Lindegren. During the years 1907–63, the following artists were responsible for designing the Physiology or Medicine diplomas: Anna Berglund, Ellen Jolin, Brita Ellstrom, Eivor Fischer, Jerk Werkmaster and Bertha Svensson-Piehl. In 1965, calligrapher Karl-Erik Forsberg designed a new Nobel diploma that excluded art work. Since 1965 the artistic decoration of the Physiology or Medicine diplomas has consisted of a gold medal in relief and a handsome calligraphic text.

**Diploma of Dario Fo, Nobel Laureate for Literature 1997**
**Artist - Bo Larsson; Calligrapher - Annika Rücker**

The Swedish Academy has always used individual designs related to each Laureate. The artists have tried to summarize something of the atmosphere and character of each author's works. Because the Prize winners are not announced until mid-October and the diplomas must be ready before December 10, the diploma artist has only a few weeks to summarize the collected works or personal attributes of each author. Aside from creating the Physiology or Medicine diplomas, Agi Lindegren also created the Literature diplomas until 1911. During 1912–18 the Literature diplomas were the work of Olle Hjortzberg, Nils Asplund and

Josef Svanlund; and during 1919–62 of Bertha Svensson (from 1938, Svensson-Piehl), with the exception of the years 1928–30 when Olle Hjortzberg was the artist. From 1963 to 1988 Gunnar Brusewitz was the man responsible for the artistic design of the Literature diplomas, and since 1989 it has been the artist Bo Larsson.

Until 1990, the Norwegian diplomas were created according to the same principles: During 1901–69, a lithograph by Gerhard Munthe and during 1970–90 a lithograph by Ørnulf Ranheimsaeter. Since 1991, different Norwegian artists have been responsible for the artistic design each year: Karl Erik Harr, Håkon Bleken, Jacob Weidemann, Anne-Lise Knoff, Ørnulf Opdahl, Jens Johannessen, Eva Arnesen, Franz Widerberg, Håvard Vikhagen and Elling Reitan.

For some years, well-known calligraphers have hand-lettered the various diplomas. The diplomas from the Royal Swedish Academy of Sciences and the Swedish Academy have been hand-lettered by Annika Rücker since 1988 and 1989, respectively, and the Physiology or Medicine diplomas by Susan Duvnas since 1990. Since 1992, the Norwegian diplomas have been hand-lettered by Inger Magnus.

**Today each Nobel diploma is a unique work of art. The Literature diploma is written on parchment, i.e. specially treated leather, using largely the same**

**Literature Laureate for 1986 Wole Soyinka's diploma box**

**technique as those of medieval book illustrators. The diplomas given to the other Laureates are produced on specially ordered handmade paper.**

The Nobel relief on the Physiology or Medicine diploma is made of leather, attached to the diploma. After extensive preparations, the bookbinder then mounts the diploma in a leather cover made of the highest quality goatskin. Nowadays the Physics diplomas are mounted in a blue leather cover, Chemistry and Physiology or Medicine in red, Econòmic Sciences in brown and Literature in the colour that the artist has chosen. In addition, the calligraphers have designed special gold monograms for each of the Laureates on the outside of the diplomas, which are also found on the boxes in which the diplomas rest. These diploma boxes are all made of gray woven paperboard, lined inside with pigskin suede. The size of the Nobel diplomas is 23 × 35 cms.

In Birger Christofferson's book *Gunnar Brusewitz,* the artist provides a description of the diploma he made for Isaac Bashevis Singer (1978): The diploma is dominated by a Star of David, whose six tips point toward characters and events in Singer's books. The pictures in the upper left portion were inspired by "The Magician of Lublin". A parrot appears there, but can also symbolize the bird that flies away with people's sins. Beneath it, a couple of rabbis with a Torah roll and ritual ram's horn. Next to it, Jacob in "The Slave", living in captivity as a cowherd. The bottom portion of the diploma is based on "Satan in Goray", with its wild ecstatic atmosphere in anticipation of Shabbetai Zvi - the false "Messiah." The flower symbolizes the recovery of Goray from devastation. And above it, New York rises as the never-realized paradise for tormented Jews. To the right, the pogroms of the Nazi era.

The book provides further examples of Nobel diploma design work. In 1984 the beloved Czech national poet Jaroslav Seifert was awarded the Nobel Prize. At 83 years old, he had a rich production behind him. The picture on the diploma was dominated by symbols of love and peace, against the backdrop of beautiful, ravaged Prague, his adored home city. In the late autumn of 1985, Gunnar Brusewitz portrayed Claude Simon's fascinating imagery, with its sharp contrasts between the gray battlefields of Flanders and surrealistic dream interpretations. In 1986, the first African

**Diploma of Camilo Jose Cela Nobel Laureate for Literature 1989 Artist - Bo Larsson; Calligrapher - Annika Rücker**

to win the Nobel Prize in Literature was Wole Soyinka of Nigeria. His diploma imparts a genuine feeling of throbbing rhythms, magic rites and the struggle for liberation.

Bo Larsson describes his art work for the 1989 Literature diploma awarded to Camilo José Cela as follows: "The black colour seemed a given: the blackness of Goya and Picasso. So I painted the whole parchment black - or almost black. A few drops of white in this black, so that the completely black pupils I would later paint would stand out clearly and intensively. The eyes would belong to Cerberus, the three-headed watchdog of Hades. I made the whites of his eyes red; he holds people in the grip of his red claws and bites them with red teeth. I then exposed the scene by sprinkling sand between the dog and the people. The sand swirls around these figures, providing a vision of movement."

# The Prize Amount in Swedish Crowns

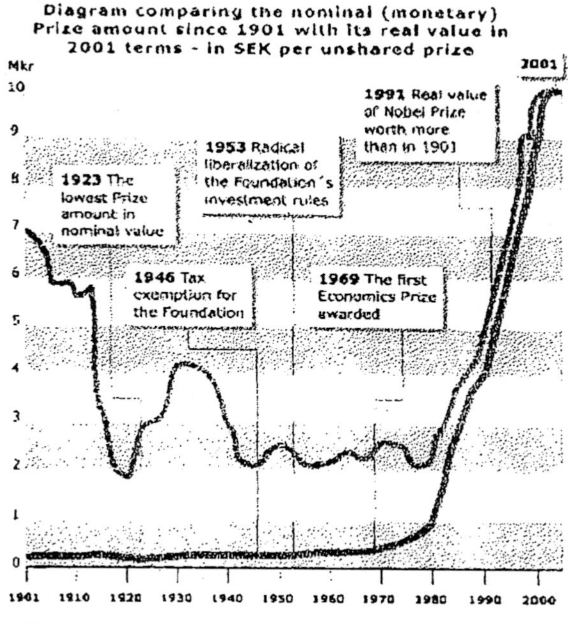


| *Year* | *Comments* | *Amount* |
|---|---|---|
| 1901 | | 150,782 |
| 1902 | | 141,847 |
| 1903 | | 141,358 |
| 1904 | | 140,859 |
| 1905 | | 138,089 |
| 1906 | | 138,536 |
| 1907 | | 138,796 |
| 1908 | | 139,800 |
| 1909 | | 139,800 |
| 1910 | | 140,703 |
| 1911 | | 140,695 |
| 1912 | | 140,476 |
| 1913 | | 143,010 |
| 1914 | | 146,900 |

| *Year* | *Comments* | *Amount* |
|---|---|---|
| 1915 | | 149,223 |
| 1916 | | 131,793 |
| 1917 | | 133,823 |
| 1918 | | 138,198 |
| 1919 | | 133,127 |
| 1920 | | 134,100 |
| 1921 | | 121,573 |
| 1922 | | 122,483 |
| 1923 | Smallest amount | 114,935 |
| 1924 | | 116,719 |
| 1925 | | 118,165 |
| 1926 | | 116,960 |
| 1927 | | 126,501 |
| 1928 | | 156,939 |
| 1929 | | 172,760 |
| 1930 | | 172,947 |
| 1931 | | 173,206 |
| 1932 | | 171,753 |
| 1933 | | 170,332 |
| 1934 | | 162,608 |
| 1935 | | 159,917 |
| 1936 | | 159,850 |
| 1937 | | 158,463 |
| 1938 | | 155,007 |
| 1939 | | 148,822 |
| 1940 | | 138,570 |
| 1941 | | 131,496 |
| 1942 | | 131,891 |
| 1943 | | 123,691 |
| 1944 | | 121,841 |
| 1945 | | 121,333 |
| 1946 | The Foundation is granted tax exemption | 121,524 |
| 1947 | | 146,115 |
| 1948 | | 159,773 |
| 1949 | | 156,290 |
| 1950 | | 164,304 |
| 1951 | | 167,612 |
| 1952 | | 171,135 |
| 1953 | The Foundation's investment rules are changed | 175,293 |

| *Year* | *Comments* | *Amount* |
|---|---|---|
| 1954 | | 181,647 |
| 1955 | | 190,214 |
| 1956 | | 200,123 |
| 1957 | | 208,629 |
| 1958 | | 214,559 |
| 1959 | | 220,678 |
| 1960 | | 225,987 |
| 1961 | | 250,233 |
| 1962 | | 257,220 |
| 1963 | | 265,000 |
| 1964 | | 273,000 |
| 1965 | | 282,000 |
| 1966 | | 300,000 |
| 1967 | | 320,000 |
| 1968 | | 350,000 |
| 1969 | The Prize in Economic Sciences is added | 375,000 |
| 1970 | | 400,000 |
| 1971 | | 450,000 |
| 1972 | | 480,000 |
| 1973 | | 510,000 |
| 1974 | | 550,000 |
| 1975 | | 630,000 |
| 1976 | | 681,000 |
| 1977 | | 700,000 |
| 1978 | | 725,000 |
| 1979 | | 800,000 |
| 1980 | | 880,000 |
| 1981 | | 1,000,000 |
| 1982 | | 1,150,000 |
| 1983 | | 1,500,000 |
| 1984 | | 1,650,000 |
| 1985 | | 1,800,000 |
| 1986 | | 2,000,000 |
| 1987 | | 2,175,000 |
| 1988 | | 2,500,000 |
| 1989 | | 3,000,000 |
| 1990 | | 4,000,000 |
| 1991 | | 6,000,000 |
| 1992 | | 6,500,000 |

| *Year* | *Comments* | *Amount* |
|---|---|---|
| 1993 | | 6,700,000 |
| 1994 | | 7,000,000 |
| 1995 | | 7,200,000 |
| 1996 | | 7,400,000 |
| 1997 | | 7,500,000 |
| 1998 | | 7,600,000 |
| 1999 | | 7,900,000 |
| 2000 | | 9,000,000 |
| 2001 | | 10,000,000 |
| 2002 | | 10,000,000 |

# List of Nobel Prize Winners in Physics

**(Year, Name, Nationality, Work)**

1901 Wilhelm Conrad Roentgen (Germany) - In recognition of the extraordinary services he has rendered by the discovery of the remarkable rays subsequently named after him.

1902 Hendrik Antoon Lorentz (Netherlands) and Pieter Zeeman (Netherlands) - In the recognition of the extraordinary services they rendered by their researches into the influence of magnetism upon radiation phenomena.

1903 Antoine Henri Becquerel (France) - In recognition of the extra ordinary services he has rendered by his discovery of spontaneous radioactivity.

Pierre Curie (France) and Marie Curie (France) - In recognition of the extraordinary services they have rendered by their joint researches on the radiation phenomena discovered by Professor Henri Becquerel.

1904 Lord (John William Strutt) Rayleigh (United Kingdom) - For his investigations of the densities of the most important gases and for his discovery of argon in connection with these studies.

1905 Philipp Eduard Anton von Lenard (Germany) - For his work on cathode rays.

1906 Joseph John Thomson (United Kingdom) - In recognition of the great merits of his theoretical and experimental investigations on the conduction of electricity by gases.

1907 Albert Abraham Michelson (USA) - For his optical precision instruments and the spectroscopic and meteorological investigations carried out with their aid.

1908 Gabriel Lippmann (France) - For his method of reproducing colours photographically based on the phenomenon of interference.

1909 Guglielmo Marconi (Italy) and Karl Ferdinand Braun (Germany) - In recognition of their contributions to the development of wireless telegraphy.

1910 Johannes Diderik vander Waals (Netherlands) - For his work on the equation of state for gases and liquids.

1911 Wilhelm Wien (Germany) - For his discoveries regarding the laws governing the radiation of heat.

1912 Nils Gustaf Dalen (Sweden) - For his invention of automatic regulators for use in conjunction with gas accumulators for illuminating lighthouses and buoys.

1913 Heike Kamerlingh Onnes (Netherlands) - For his investigations on the properties of matter at low temperatures which led, inter alia, to the production of liquid helium.

1914 Max von Laue (Germany) - For his discovery of the diffraction of X-rays by crystals.

1915 Sir William Henry Bragg (United Kingdom) and William Lawrence Bragg (United Kingdom) - For their services in the analysis of crystal structure by means of X-rays.

1916 Not Awarded.

1917 Charles Glover Barkla (United Kingdom) - For his discovery of the characteristic Roéntgen radiation of the elements.

1918 Max Karl Ernst Ludwig Planck (Germany) - In recognition of the services he rendered to the advancement of physics by his discovery of energy quanta.

1919 Johannes Stark (Germany) - For his discovery of the Doppler effect in canal rays and the splitting of spectral lines in electric fields.

1920 Charles Edouard Guillaume (Switzerland) - In recognition of the service he has rendered to precision measurements in physics by his discovery of anomalies in nickel steel alloys.

1921 Albert Einstein (Germany and Switzerland) - For his services to theoretical physics, and especially for his discovery of the laws of the photoelectric effect.

1922 Niels Henrik David Bohr (Denmark) - For his services in the investigation of the structure of atoms and of the radiation emanating from them.

1923 Robert Andrews Millikan (USA) - For his work on the elementary charge of electricity and on the photoelectric effect.

1924 Karl Manne Georg Siegbahn (Sweden) - For his discoveries and research in the field of X-ray spectroscopy.

1925 James Franck (Germany) and Gustav Ludwig Hertz (Germany) - For their discovery of the laws governing the impact of an electron upon an atom.

1926 Jean Baptiste Perrin (France) - For his work on the discontinuous structure of matter, and especially for his discovery of sedimentation equilibrium.

1927 Arthur Holly Compton (USA) - For his discovery of the effect named after him.

Charles Thomson Rees Wilson (United Kingdom) - For his method of making the paths of electrically charged particles visible by condensation of vapour.

1928 Owen Willans Richardson (United Kingdom) - For his work on the thermionic phenomenon and especially for the discovery of the law named after him.

1929 Prince Louis-Victor Pierre Raymond de Broglie (France) - For his discovery of the wave nature of electrons.

1930 Sir Chandrasekhara Venkata Raman (India) - For his work on the scattering of light and for the discovery of the effect named after him.

1931 Not Awarded.

1932 Werner Karl Heisenberg (Germany) - For the creation of quantum mechanics the application of which has, inter alia, led to the discovery of the allotropic forms of hydrogen.

1933 Erwin Schrodinger (Austria) and Paul Adrien Maurice Dirac (United Kingdom) - For the discovery of new productive forms of atomic theory.

1934 Not Awarded.

1935 James Chadwick (UnitedKingdom) - For the discovery of the neutron.

1936 Victor Franz Hess (Austria) - For his discovery of cosmic radiation.

Carl David Anderson (USA) - For his discovery of positron.

1937 Clinton Joseph Davisson (USA) and George Paget Thomson (United Kingdom) - For their experimental discovery of the diffraction of electron by crystals.

1938 Enrico Fermi (Italy) - For his demonstrations of the existence of new radioactive elements produced by neutron irradiation, and for his related discovery of nuclear reactions brought about by slow neutrons.

1939 Ernest Orlando Lawrence (USA) - For the invention and development of the cyclotron for results obtained with it, especially with regard to artificial radioactive elements.

1940 Not Awarded.

1941 Not Awarded.

1942 Not Awarded.

1943 Otto Stern (USA) - For his contribution to the development of the molecular ray method and his discovery of the magnetic moment of the proton.

1944 Isidor Isaac Rabi (USA) - For his resonance method for recording the magnetic properties of atomic nuclei.

1945 Wolfgang Pauli (Austria) - For the discovery of the Exclusion Principle, also called the Pauli Principle.

1946 Percy Williams Bridgman (USA) - For the invention of an apparatus to produce extremely high pressures, and for the discoveries he made there with in the field of high pressure physics.

1947 Sir Edward Victor Appleton (United Kingdom) - For his investigations of the physics of the upper atmosphere especially for the discovery of the so-called Appleton Layer.

1948 Patrick Maynard Stuart Blackett (United Kingdom) - For his development of the Wilson cloud chamber method, and his discoveries theirwith in the fields of nuclear physics and cosmic radiation.

1949 Hideki Yukawa (Japan) - For his prediction of the existence of mesons on the basis of theoretical work on nuclear forces.

1950 Cecil Frank Powell (United Kingdom) - For his development of the photographic method of studying nuclear processes and his discoveries regarding mesons made with this methods.

1951 Sir John Douglas Cockcroft (United Kingdom) and Ernest Thomas Sinton Walton (Ireland) - For their pioneer work on the transmutation of atomic nuclei by artificially accelerated atomic particles.

1952 Felix Bloch (USA) and Edward Mills Purcell (USA) - For their development of new methods for nuclear magnetic precision measurements and discoveries in connection therewith.

1953 Frits (Frederik) Zernike (Netherlands) - For his demonstration of the phase contrast method, especially for his invention of the phase contrast microscope.

1954 Max Born (United Kingdom) - For his fundamental research in quantum mechanics, especially for his statistical interpretation of the wave function.

Walther Bothe (Federal Republic of Germany) - For the coincidence method and his discoveries made therewith.

1955 Willis Eugene Lamb (USA) - For his discoveries concerning the fine structure of the hydrogen spectrum.

Polykarp Kusch (USA) - For his precision determination of the magnetic moment of the electron.

1956 William Bradford Shockley (USA), John Bardeen (USA) and Walter-Houser Brattain (USA) - For their researches on semiconductors and their discovery of the transistor effect.

1957 Chen Ning Yang (China) and Tsung-Dao Lee (China) - For their penetrating investigation of the so-called parity laws which has led to important discoveries regarding the elementary particles.

1958 Pavel Alekseyevich Cherenkov (USSR), Il'ja Mikhailovich Frank (USSR) and Igor Yevgenyevich Tamm (USSR) - For the discovery and the interpretation of the Cherenkov effect.

1959 Emilio Gino Segre (USA) and Owen Chamberlain (USA) - For their discovery of the antiproton.

1960 Donald Arthur Glaser (USA) - For the invention of bubble chamber.

1961 Robert Hofstadter (USA) - For his pioneering studies of electron scattering in atomic nuclei and for his thereby achieved discoveries concerning the structure of the nucleons.

Rudolf Ludwig Mossbauer (Federal Republic of Germany) - For his researches concerning the resonance absorption of gamma radiation and his discovery in this connection of the effect which bears his name.

1962 Lev Davidovich Landau (USSR) - For his pioneering theories for condensed matter, especially liquid helium.

1963 Eugene Paul Wigner (USA) - For his contribution to the theory of the atomic nucleus and the elementary particles, particularly through the discovery and application of fundamental symmetry principles.

Maria Goeppert-Mayer (USA) and J. Hans D. Jensen (Federal Republic of Germany) - For their discoveries concerning nuclear shell structure.

1964 Charles Hard Townes (USA), Nicolay Gennadiyevich Basov (USSR) and Aleksandr Mikhailovich Prokhorov (USSR) - For fundamental work in the field of quantum electronics, which has led to the construction of oscillators and amplifiers based on maser-laser principle.

1965 Sin-Itiro Tomonaga (Japan), Julian Schwinger (USA) and Richard P. Feynman (USA) - For their fundamental work in quantum electrodynamics, with deep ploughing consequences for the physics of elementary particles.

1966 Alfred Kastler (France) - For the discovery and development of optical methods for studying Hertzian resonances in atoms.

1967 Hans Albrecht Bethe (USA) - For his contribution to the theory of nuclear reactions, especially his discoveries concerning the energy production in stars.

1968 Luis Walter Alvarez (USA) - For his decisive contribution to elementary particle physics, in particular the discovery of a large number of resonance states, made possible through his development of the technique of using hydrogen bubble chamber and data analysis.

1969 Murray Gell-Mann (USA) - For his contributions and discoveries concerning the classification of elementary particle and their interactions.

1970 Hannes Olof Gsta Alfven (Sweden) - For fundamental work and discoveries in magneto-hydrodynamics with fruitful applications in different parts of plasma physics.

Louis Eugene Felix Neel (France) - For fundamental work and discoveries concerning antiferromagnetism and ferrimagnetism which have led to the important applications in solid state physics.

1971 Dennis Gabor (United Kingdom) - For his invention and development of the holographic method.

1972 John Bardeen (USA), Leon Neil Cooper (USA), John Robert Schrieffer (USA) - For their jointly developed theory of superconductivity, usually called the BCS-theory.

1973 Leo Esaki (Japan) and Ivar Giaever (USA) - For their experimental discoveries regarding tunneling phenomena in semiconductors and superconductors respectively.

Brian David Josephson (United Kingdom) - For his theoretical predictions of the properties of a supercurrent through a tunnel barrier, in particular those phenomena which are generally known as the Josephson effects.

1974 Martin Ryle (United Kingdom) and Antony Hewish (United Kingdom) - For their pioneering research in radio astrophysics: Ryle for his observations and inventions, in particular of the aperture synthesis technique, and Hewish for his decisive role in the discovery of pulsars.

1975 Aage Niels Bohr (Denmark), Ben Roy Mottelson (Denmark) and Leo James Rainwater (USA) - For the discovery of the connection between collective motion and particle motion in atomic nuclei and the development of the theory of the structure of the atomic nucleus based on this connection.

1976 Burton Richter (USA) and Samuel Chao Chung Ting (USA) - For their pioneering work in the discovery of a heavy elementary particle of a new kind.

1977 Philip Warren Anderson (USA), Sir Nevill Francis Mott (United Kingdom) and John Hasbrouck van Vleck (USA) - For their fundamental theoretical investigations of the electronic structure of magnetic and disordered systems.

1978 Pyotr Leonidovich Kapitza (USSR) - For his basic inventions and discoveries in the area of low-temperature physics.

Arno Allan Penzias (USA) and Robert Woodrow Wilson (USA) - For their discovery of cosmic microwave background radiation.

1979 Sheldon Lee Glashow (USA), Abdus Salam (Pakistan), Steven Weinberg (USA) - For their contributions to the theory of the unified weak and electromagnetic interaction between elementary particles, including, inter alia, the prediction of the weak neutral current.

1980 James Watson Cronin (USA), Val Logsdon Fitch (USA) - For the discovery of violations of fundamental symmetry principles in the decay of neutral K-mesons.

1981 Nicolaas Bloembergen (USA), Arthur Leonard Schawlow (USA) - For their contribution to the development of laser spectroscopy.

Kai M. Siegbahn (Sweden) - For his contribution to the development of high resolution electron spectroscopy.

1982 Kenneth G. Wilson (USA) - For his theory for critical phenomena in connection with phase transitions.

1983 Subramanyan Chandrasekhar (USA) - For his theoretical studies of the physical processes of importance to the structure and evolution of the stars.

William Alfred Fowler (USA) - For his theoretical and experimental studies of the nuclear reactions of importance in the formation of the chemical elements in the universe.

1984 Carlo Rubbia (Italy), Simon vander Meer (Netherlands) - For their decisive contributions to the large project, which led to the discovery of the field particles W and Z, communicators of weak interaction.

1985 Klaus von Klitzing (Federal Republic of Germany) - For the discovery of the quantized hall effect.

1986 Ernst Ruska (Federal Republic of Germany) - For his fundamental work in electron optics, and for the design of the first electron microscope.

Gerd Binnig (Federal Republic of Germany), Heinrich Rohrer (Switzerland) - For their design of the scanning tunneling microscope.

1987 J. Georg Bednorz (Federal Republic of Germany) and K. Alexander Muller (Switzerland) - For their important break-through in the discovery of super conductivity in ceramic materials.

1988 Leon M. Lederman (USA), Melvin Schwartz (USA) and Jack Steinberger (USA) - For the neutrino beam method and the demonstration of the doublet structure of the leptons through the discovery of the muon neutrino.

1989 Norman F. Ramsey (USA) - For the invention of the separated oscillatory fields method and its use in the hydrogen maser and other atomic clocks.

Hans G. Dehmelt (USA) and Wolfgang Paul (Federal Republic of Germany) - For the development of the ion trap technique.

1990 Jerome I. Friedman (USA), Henry W. Kendall (USA) and Richard E. Taylor (Canada) - For their pioneering investigations concerning deep inelastic scattering of electrons on protons and bound neutrons, which have been of essential importance for the development of the quark model in particle physics.

1991 Pierre-Gilles de Gennes (France) - For discovering that methods developed for studying order phenomena in simple systems can be generalized to more complex forms of matter, in particular to liquid crystals and polymers.

1992 Georges Charpak (France) - For his invention and development of particle detectors, in particular the multiwire proportional chamber.

1993 Russell A. Hulse (USA) and Joseph H. Taylor Jr. (USA) - For the discovery of a new type of pulsar, a discovery that has opened up new possibilities for the study of gravitation.

1994 Bertram N. Brockhouse (Canada) - For the development of neutron spectroscopy.

Clifford G. Shull (USA) - For the development of the neutron diffraction technique.

"For pioneering contributions to the development of neutron scattering techniques for studies of condensed matter".

1995 Martin L. Perl (USA) - For the discovery of the tau lepton.

Frederick Reines (USA) - For the detection of the neutrino.

"For pioneering experimental contributions to lepton physics."

1996 David M. Lee (USA), Douglas D. Osheroff (USA) and Robert C. Richardson (USA) - For their discovery of superfluidity in Helium-3

1997 Steven Chu (USA), Claude Cohen-Tannoudji (France) and William D. Phillips (USA) - For development of methods to cool and trap atoms with laser light.

1998 Robert B. Laughlin (USA), Horst L. Stormer (Federal Republic of Germany) and Daniel C. Tsui (USA) - For their discovery of a new form of quantum fluid with fractionally charged excitations.

1999 Gerardus't Hooft (Netherlands) and Martinus J. G. Veltman (Netherlands) - For elucidating the quantum structure of electroweak interactions in physics.

2000 Zhores I. Alferov (Russia) and Herbert Kroemer (Federal Republic of Germany) - For developing semiconductor heterostructures used in high speed and opto-electronics.

Jack S. Kilby (USA) - For his part in the invention of the integrated circuits.

"For basic work on information and communication technology."

2001 Eric A. Cornell (USA) Wolfgang Ketterle (Federal Republic of Germany) and Carl E. Wieman (USA) - For the achievement of Bose-Einstein condensation in dilute gases of alkali atoms, and for early fundamental studies of the properties of the condensates.

2002 Raymond Davis Jr. (USA), Masatoshi Koshiba (Japan) - For pioneering contributions to astrophysics, in particular for the detection of cosmic neutrinos.

Riccardo Giacconi (USA) - For pioneering contributions to astrophysics, which have led to the discovery of cosmic X-ray sources.

# LIST OF NOBEL PRIZE WINNERS IN PHYSICS

**(Year, Name, Nationality)**

1901 Wilhelm C. Roentgen (Germany)
1902 Hendrik A. Lorentz (Netherlands)
Pieter Zeeman (Netherlands)
1903 Pierre Curie (France)
Marie Curie (France)
Antoine H. Becquerel (France)
1904 John W. Rayleigh (United Kingdom)
1905 Philipp Lenard (Germany)
1906 Sir Joseph John Thomson (United Kingdom)
1907 Albert A. Michelson (United States)
1908 Gabriel Lippmann (France)
1909 Karl F. Braun (Germany)
Marchese G. Marconi (Italy)
1910 Johannes D. van der Waals (Netherlands)
1911 Wilhelm Wien (Germany)
1912 Nils Gustaf Dalén (Sweden)
1913 Heike Kamerlingh Onnes (Netherlands)
1914 Max von Laue (Germany)
1915 Sir William Henry Bragg (United Kingdom)
Sir William L. Bragg (United Kingdom)
1916 Not awarded
1917 Charles G. Barkla (United Kingdom)
1918 Max Karl E.L. Planck (Germany)
1919 Johannes Stark (Germany)
1920 Charles E. Guillaume (France)
1921 Albert Einstein (United States)
1922 Niels Henrik D. Bohr (Denmark)
1923 Robert A. Millikan (United States)
1924 Karl M.G. Siegbahn (Sweden)
1925 James Franck (Germany)
Gustav Hertz (Germany)
1926 Jean Baptiste Perrin (France)
1927 Arthur H. Compton (United States)
Charles T. Wilson (United Kingdom)
1928 Sir Owen Richardson (United Kingdom)
1929 Louis Victor de Broglie (France)
1930 Sir Chandrasekhara Raman (India)
1931 Not awarded

1932 Werner Heisenberg (Germany)
1933 Paul Adrien M. Dirac (United Kingdom)
Erwin Schrödinger (Austria)
1934 Not awarded
1935 Sir James Chadwick (United Kingdom)
1936 Carl D. Anderson (United States)
Victor F. Hess (Austria)
1937 Clinton J. Davisson (United States)
Sir George P. Thomson (United Kingdom)
1938 Enrico Fermi (Italy)
1939 Ernest O. Lawrence (United States)
1940 Not awarded
1941 Not awarded
1942 Not awarded
1943 Otto Stern (United States)
1944 Isidor I. Rabi (United States)
1945 Wolfgang Pauli (United States)
1946 Percy W. Bridgman (United States)
1947 Sir Edward V. Appleton (United Kingdom)
1948 Patrick M.S. Blackett (United Kingdom)
1949 Hideki Yukawa (Japan)
1950 Cecil F. Powell (United Kingdom)
1951 Sir John D. Cockcroft (United Kingdom)
Ernest T.S. Walton (Ireland)
1952 Felix Bloch (United States)
Edward M. Purçell (United States)
1953 Frits Zernike (Netherlands)
1954 Max Born (West Germany)
Walther Bothe (East Germany)
1955 Polykarp Kusch (United States)
Willis Eugene Lamb, Jr. (United States)
1956 William B. Shockley (United States)
Walter H. Brattain (United States)
John Bardeen (United States)
1957 Tsung-Dao Lee (China-United States)
Chen Ning Yang (United States)
1958 Pavel A. Cherenkov (Union of Soviet Socialist Republics)
Il'ja M. Frank (Union of Soviet Socialist Republics)
Igor Y. Tamm (Union of Soviet Socialist Republics)
1959 Owen Chamberlain (United States)
Emilio G. Segré (United States)

1960 Donald A. Glaser (United States)
1961 Robert Hofstadter (United States)
Rudolf L. Mössbauer (West Germany)
1962 Lev. D. Landau (Union of Soviet Socialist Republics)
1963 Eugene P. Wigner (United States)
Maria G. Mayer (United States)
Johannes Hans D. Jensen (West Germany)
1964 Nicolay G. Basov (Union of Soviet Socialist Republics)
Aleksandr Mikhailovich Prokhorov (Union of Soviet Socialist Republics)
Charles H. Townes (United States)
1965 Richard P. Feynman (United States)
Julian S. Schwinger (United States)
Sin-Itiro Tomonaga (Japan)
1966 Alfred Kastler (France)
1967 Hans A. Bethe (United States)
1968 Luis W. Alvarez (United States)
1969 Murray Gell-Mann (United States)
1970 Louis Eugène F. Néel (France)
Hannes O.G. Alfvén (Sweden)
1971 Dennis Gabor (United Kingdom)
1972 John Bardeen (United States)
Leon N. Cooper (United States)
John R. Schrieffer (United States)
1973 Ivar Giaever (United States)
Leo Esaki (Japan)
Brian D. Josephson (United Kingdom)
1974 Sir Martin Ryle (United Kingdom)
Antony Hewish (United Kingdom)
1975 Aage N. Bohr (Denmark)
Ben R. Mottelson (Denmark)
James Rainwater (United States)
1976 Burton Richter (United States)
Samuel C.C. Ting (United States)
1977 John H. Van Vleck (United States)
Philip W. Anderson (United States)
Sir Nevill F. Mott (United Kingdom)
1978 Arno A. Penzias (United States)
Robert W. Wilson (United States)
Pyotr Leonidovich Kapitza (Union of Soviet Socialist Republics)

1979 Steven Weinberg (United States)
Sheldon L. Glashow (United States)
Abdus Salam (Pakistan)
1980 James W. Cronin (United States)
Val L. Fitch (United States)
1981 Nicolaas Bloembergen (United States)
Arthur Leonard Schawlow (United States)
Kai Siegbahn (Sweden)
1982 Kenneth G. Wilson (United States)
1983 Subrahmanyan Chandrasekhar (United States)
William A. Fowler (United States)
1984 Carlo Rubbia (Italy)
Simon van der Meer (Netherlands)
1985 Klaus von Klitzing (West Germany)
1986 Ernest Ruska (West Germany)
Gerd Binnig (West Germany)
Heinrich Rohrer (Switzerland)
1987 K. Alex Muller (Switzerland)
J. Georg Bednorz (Switzerland)
1988 Leon Max Lederman (United States)
Melvin Schwartz (United States)
Jack Steinberger (United States)
1989 Hans G. Dehmelt (United States)
Wolfgang Paul (West Germany)
Norman F. Ramsey (United States)
1990 Richard E. Taylor (Canada)
Jerome I. Friedman (United States)
Henry W. Kendall (United States)
1991 Pierre Gilles de Gennes (France)
1992 Georges Charpak (France)
1993 Joseph H. Taylor (United States)
Russell A. Hulse (United States)
1994 Clifford G. Shull (United States)
Bertram N. Brockhouse (Canada)
1995 Martin L. Perl (United States)
Frederick Reines (United States)
1996 David M. Lee (United States)
Douglas D. Osheroff (United States)
Robert C. Richardson (United States)

| | |
|---|---|
| 1997 | Steven Chu (United States) |
| | Claude Cohen-Tannoudji (France) |
| | William D. Phillips (United States) |
| 1998 | Robert Laughlin (United States) |
| | Daniel Tsui (United States) |
| | Horst Störmer (Germany) |
| 1999 | Gerardus 't Hooft (Netherlands) |
| | Martinus J.G. Veltman (Netherlands) |
| 2000 | Zhores I. Alferov (Russia) |
| | Herbert Kroemer (Federal Republic of Germany) |
| | Jack S. Kilby (United States) |
| 2001 | Eric A. Cornell (United States) |
| | Wolfgang Ketterle (Federal Republic of Germany) |
| | Carl E. Wieman (United States) |
| 2002 | Raymond Davis (United States) |
| | Masatoshi Koshiba (Japan) |
| | Riccardo Giacconi (United States) |

# ILLUSTRATIONS

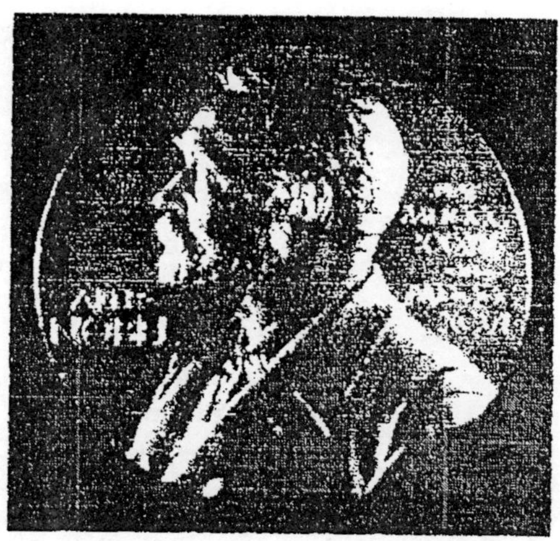

The obverse side of the Nobel Prize medals for Physics, Chemistry, Physiology or Medicine, and Literature.

The reverse side of the Nobel Prize medal awarded for both Physics and Chemistry.

The reverse side of the Nobel Prize medal for Physiology or Medicine.

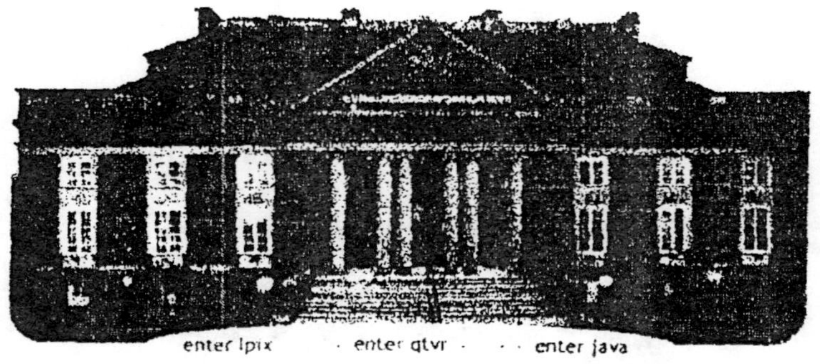


Virtual Tour of the Royal Swedish Academy of Sciences

Nobel Stamps

In 1946 the first stamp portraying Alfred Nobel was released. After that a few set of stamps have commemorated institutions and buildings associated with the Nobel Prize. It was not until fifteen years later, however, that Sweden Post Stamps decided to commemorate Nobel Prize winners in a yearly edition of stamps. In 1961 the Nobel annual series was established.

Part
II

*1901 Wilhelm C. Roentgen (Germany)*
*1902 Hendrik A. Lorentz (Netherlands)*
*Pieter Zeeman (Netherlands)*
*1903 Pierre Curie (France)*
*Marie Curie (France)*
*Antoine H. Becquerel (France)*
*1904 John W. Rayleigh (United Kingdom)*
*1905 Philipp Lenard (Germany)*
*1906 Sir Joseph John Thomson (United Kingdom)*
*1907 Albert A. Michelson (United States)*
*1908 Gabriel Lippmann (France)*
*1909 Karl F. Braun (Germany)*
*Marchese G. Marconi (Italy)*
*1910 Johannes D. van der Waals (Netherlands)*
*1911 Wilhelm Wien (Germany)*
*1912 Nils Gustaf Dalén (Sweden)*
*1913 Heike Kamerlingh Onnes (Netherlands)*
*1914 Max von Laue (Germany)*
*1915 Sir William Henry Bragg (United Kingdom)*
*Sir William L. Bragg (United Kingdom)*
*1916 Not awarded*
*1917 Charles G. Barkla (United Kingdom)*
*1918 Max Karl E.L. Planck (Germany)*
*1919 Johannes Stark (Germany)*
*1920 Charles E. Guillaume (France)*
*1921 Albert Einstein (United States)*
*1922 Niels Henrik D. Bohr (Denmark)*
*1923 Robert A. Millikan (United States)*
*1924 Karl M.G. Siegbahn (Sweden)*
*1925 James Franck (Germany)*
*Gustav Hertz (Germany)*
*1926 Jean Baptiste Perrin (France)*
*1927 Arthur H. Compton (United States)*
*Charles T. Wilson (United Kingdom)*
*1928 Sir Owen Richardson (United Kingdom)*
*1929 Louis Victor de Broglie (France)*
*1930 Sir Chandrasekhara Raman (India)*
*1931 Not awarded*
*1932 Werner Heisenberg (Germany)*
*1933 Paul Adrien M. Dirac (United Kingdom)*
*Erwin Schrödinger (Austria)*
*1934 Not awarded*
*1935 Sir James Chadwick (United Kingdom)*
*1936 Carl D. Anderson (United States)*
*Victor F. Hess (Austria)*

Wilhelm Roentgen

*"German physicist Wilhelm Roentgen won the 1901 Nobel Prize in physics. Roentgen, who was the first Nobel Laureate in physics, won the award for his discovery of a type of short-wave radiation popularly known as X-rays."*

# WILHELM CONRAD ROENTGEN

**Roentgen, Wilhelm Conrad** (1845-1923), German physicist, discoverer of X-rays, and winner of the first Nobel Prize in physics. Roentgen's discovery of X-rays was a momentous advance for physics and medicine and earned him the 1901 Nobel Prize in physics.

Roentgen was born in Lennep, Germany, and grew up in the Netherlands. He earned a mechanical engineering degree at the Federal Institute of Technology in Zürich, Switzerland, in 1868 and a Ph.D. degree in physics at the University of Zürich in 1869. Roentgen worked as a laboratory assistant at the University of Würzburg in Germany from 1868 to 1872 and at the University of Strasbourg in Germany from 1872 to 1874. He began teaching physics in 1874, starting at the University of Strasbourg, then moving to the Agricultural Academy in Hohenheim, Germany, in 1875, and back to Strasbourg in 1876. In 1879 he became a professor of physics at the University of Giessen in Germany, where he remained until 1888, when he became professor of physics and director of the Physical Institute at the University of Würzburg. He accepted a position as professor of physics and director of the Physical Science Institute at the University of Munich in 1899 and taught there until his retirement in 1920.

Roentgen made an accidental discovery November 8, 1895, while investigating emissions from a Crookes tube (a glass vacuum tube with electrodes at either end). The emissions Roentgen was looking for are called cathode rays, composed of high-speed electrons that come off the negative electrode when voltage is applied to the electrodes of a Crookes tube. Cathode rays cause the vacuum tube to glow when the vacuum is strong enough and enough voltage is applied; it was this glow Roentgen was watching when he made his discovery. Cathode rays are weak-they cannot pass through glass (scientists use aluminum *windows* in a vacuum tube when they want to study cathode rays outside the tube) or an ordinary piece of cardboard-but they do excite barium platinocyanide molecules and cause surfaces painted with barium platinocyanide to glow.

Roentgen was using a Crookes tube without an aluminum window. and he had surrounded the tube in black cardboard to better see the tube glow. Therefore, when he noticed a glow coming from a screen painted with barium platinocyanide that was some distance away, he knew cathode rays could not be the cause, because cathode rays could not have gone through either the glass of the tube or the cardboard. Roentgen did more tests to verify that the Crookes tube was the source of the emissions that made the screen glow. He inferred that these emissions were present in all experiments like his, but that he was the first to notice them because no one else had set up conditions like his experiment: a nearby barium platinocyanide-painted screen and an arrangement that suppressed the known emissions.

Roentgen called the newly discovered emissions X-rays (X is the symbol for the unknown in mathematics) and worked to document them further. He found a photographic plate in a drawer of a desk in the same room as the Crookes tube and noticed it had been exposed. When he developed it and found an image of a key that had been on the desk top, he realized the X-rays had passed easily through the wood of the desk, but to a lesser degree through the metal of the key. Using a barium platinocyanide-treated screen and a Crookes tube, Roentgen produced an image of a lead disk-and the bones of his fingers holding the disk.

His experiments showed that X-rays pass through different materials to different degrees. Release of his findings caused worldwide excitement and speculation about X-rays, also called Roentgen rays. Medical applications in diagnosis began immediately. The hazard of burns from prolonged exposure soon became evident; the risk of cancer was realized later. X-rays came to be used in medical treatment, dental examinations, industrial inspections of metal work, and many other areas.

Roentgen explained many fundamental laws of X-rays in three scientific papers, published in 1895, 1896, and 1897, but left the continuation of X-ray research to other physicists. His fame for the discovery of X-rays overshadowed his important studies of gases and crystalline substances, conducted between 1876 and 1895.

Hendrik Antoon Lorentz

*"Dutch physicist Hendrik A. Lorentz won the 1902 Nobel Prize in physics. His research helped explain the Zeeman effect, the splitting of a spectral line when the light source is in a magnetic field."*

# Hendrik Antoon Lorentz

**Lorentz, Hendrik Antoon** (1853-1928), Dutch physicist and Nobel Laureate. Lorentz was born in Arnhem and educated at the Leiden University, where he became professor of mathematical physics in 1878. He developed the electromagnetic theory of light and the electron theory of matter and formulated a consistent theory of electricity, magnetism, and light. With the Irish physicist George Francis FitzGerald, he formulated a theory on the change in shape of a body resulting from its motion; the effect, known as the Lorentz-FitzGerald contraction, was one of several important contributions that Lorentz made to the development of the theory of relativity. For his explanation of the phenomenon known as the Zeeman effect, Lorentz shared the 1902 Nobel Prize in physics with the Dutch physicist Pieter Zeeman.

**Pieter Zeeman**

*"Dutch physicist Pieter Zeeman won the 1902 Nobel Prize in physics. His work explored the influences of magnetism on radiation and light, the latter resulting in a scientific understanding of what is now called the Zeeman effect."*

# PIETER ZEEMAN

**Zeeman, Pieter** (1865-1943), Dutch physicist and Nobel Laureate. He is best known for his discovery of the Zeeman effect. Zeeman was born in Zonnemaire and educated at the Leiden University, where he taught until he became professor of physics at the University of Amsterdam in 1900. In 1896 Zeeman discovered that the spectral lines of a light source subjected to a strong magnetic field were split into several components, each of which was polarized. This phenomenon, known as the Zeeman effect, confirmed the electromagnetic theory of light. Zeeman shared the 1902 Nobel Prize in physics with the Dutch physicist Hendrik Antoon Lorentz for their joint research into the influence of magnetism upon radiation.

Pierre Curie

*"French chemist Pierre Curie made several important contributions to the study of magnetism in his early life. He and his wife, Marie Curie, shared the 1903 Nobel Prize in physics for their discovery of two new elements–radium and polonium."*

# PIERRE CURIE

**Curie, Pierre** (1859-1906), French physicist and Nobel laureate, best known for the work on radioactivity that he did with his wife, Marie Curie. In radioactive materials the atoms break down spontaneously, releasing radiation in the form of energy and subatomic particles. Pierre Curie also worked on important topics in the structure of crystals and helped discover the piezoelectric effect in crystals-a property of producing electrical voltages when they are compressed.

Pierre Curie was born in Paris and educated at home by his parents. He studied physics at the University of Paris, earning a bachelor's degree in 1875. He became an assistant teacher at the University of Paris in 1878 and turned his research to crystallography. In 1880 Pierre and his brother Jacques Curie discovered that some crystals developed positive electrical charges at one end and negative electrical charges at the other when the crystals were compressed. These crystals also change shape when exposed to electric voltage. The Curies called this effect the piezoelectric effect.

In 1894 Pierre Curie and Marie Sklodowska were introduced to one another. Their mutual devotion to scientific study led to their marriage in 1895. The same year, Pierre earned a doctoral degree in physics from the University of Paris for his research on magnetism. He showed that magnetic materials made of iron compounds lose their magnetic properties if heated beyond a certain temperature. This temperature, different for every material, is now called the Curie point. Until the mid-1890s most of Curie's research was on magnetism and on crystals.

From 1895 on, the Curies worked on radioactivity. In 1898 they isolated the element radium from pitchblende, a radioactive mineral (mineral whose atoms spontaneously emit energy and subatomic particles) that also contains uranium. They shared the 1903 Nobel Prize in physics with French physicist Antoine Henri Becquerel for their work on radioactivity. Pierre became a professor of physics at the University of Paris in 1904. Shortly after accepting the appointment, he was struck and killed by a horse-drawn carriage.

Marie Curie

*"Marie Curie was the first woman to win the Nobel Prize and also the first person to win the Nobel Prize twice one in physics in 1903 and other in chemistry in 1911. Curie coined the term "radioactive" to describe the uranium emissions she observed in early experiments. With her husband, she later discovered the elements polonium and radium. A dedicated and respected physicist, her brilliant work with radioactivity eventually cost her life, she died from overexposure to radiation."*

# Marie Curie

**Curie, Marie** (1867-1934), Polish-born French chemist who, with her husband Pierre Curie, was an early investigator of radioactivity. Radioactivity is the spontaneous decay of certain elements into other elements and energy. The Curies shared the 1903 Nobel Prize in physics with French physicist Antoine Henri Becquerel for fundamental research on radioactivity. Marie Curie went on to study the chemistry and medical applications of radium. She was awarded the 1911 Nobel Prize in chemistry in recognition of her work in discovering radium and polonium and in isolating radium

Marie Curie's maiden name was Maria Sk³odowska, and her nickname while growing up was Manya. She was born in Warsaw at a time when Poland was under Russian domination after the unsuccessful revolt of 1863. Her parents were teachers, but soon after Manya (their fifth child) was born, they lost their teaching posts and had to take in boarders. Their young daughter worked long hours helping with the meals, but she nevertheless won a medal for excellence at the local high school, where the examinations and some classes were held in Russian. No higher education was available to women in Poland at that time, so Manya took a job as a governess. She sent part of her earnings to Paris to help pay for her older sister's medical studies. Her sister qualified as a doctor and married a fellow doctor in 1891. Manya went to join them in Paris, changing her name to Marie. She entered the Sorbonne (now the Universities of Paris) and studied physics and mathematics, graduating at the top of her class. In 1894 she met the French chemist Pierre Curie, and they were married the following year.

From 1896 the Curies worked together on radioactivity, building on the results of German physicist Wilhelm Roentgen (who had discovered X rays) and Henri Becquerel (who had discovered that uranium salts emit similar radiation). Marie Curie discovered that the metallic element thorium also emits radiation and found that the mineral pitchblende emitted even more radiation than its uranium and thorium content could cause. The Curies then carried out an exhaustive search for the substance that could be producing the radioactivity. They processed an enormous amount of pitchblende, separating it into its chemical components. In July 1898 the Curies announced the discovery of the element polonium, followed in December of that year with the discovery of the element radium. They eventually prepared 1 g (0.04 oz) of pure radium chloride from 8 metric tons of waste pitchblende from Austria. They also established that beta rays (now known to consist of electrons) are negatively charged particles.

In 1906 Marie took over Pierre Curie's post at the Sorbonne when he was run down and killed by a horse-drawn carriage. She became the first woman to teach there, and she concentrated all her energies into research and caring for her daughters. The Curies' older daughter, Irene, later married Frédéric Joliot and became a famous scientist. In 1910 Marie worked with French chemist André Debierne to isolate pure radium metal. In 1914 the University of Paris built the Institut du Radium (now the Institut Curie) to provide laboratory space for research on radioactive materials.

At the outbreak of World War I in 1914, Marie Curie helped to equip ambulances with X-ray equipment, which she drove to the front lines. The International Red Cross made her head of its Radiological Service. She and her colleagues at the Institut du Radium held courses for medical orderlies and doctors, teaching them how to use the new technique. By the late 1920's her health began to deteriorate: Continued exposure to high-energy radiation had given her leukemia. She entered a sanatorium at Haute

Savoie and died there on July 4, 1934, a few months after her daughter and son-in-law, the Joliot-Curies, announced the discovery of artificial radioactivity.

Throughout much of her life Marie Curie was poor, and she and her fellow scientists carried out much of their work extracting radium under primitive conditions. The Curies refused to patent any of their discoveries, wanting them to benefit everyone freely. The Nobel Prize money and other financial rewards were used to finance further research. One of the outstanding applications of their work has been the use of radiation to treat cancer, one form of which cost Marie Curie her life.

Antoine Henri Becquerel

*"French physicist Antoine Henri Becquerel won the Nobel Prize in physics in 1903. Becquerel discovered radioactivity in uranium."*

# ANTOINE HENRI BECQUEREL

**Becquerel, Antoine Henri** (1852-1908), French physicist and Nobel laureate, who discovered radioactivity in uranium. He was the son of Alexandre Becquerel, who studied light and phosphorescence and invented the phosphoroscope, and grandson of Antoine César Becquerel, one of the founders of electrochemistry.

Born in Paris, Becquerel became professor of physics at the Museum of Natural History in 1892 and at the Polytechnical School in 1895. In 1896 he accidentally discovered the phenomenon of radioactivity in the course of his research on fluorescence. After placing uranium salts on a photographic plate in a dark area, Becquerel found that the plate had become blackened. This proved that uranium must give off its own energy, which later became known as radiation.

Becquerel also conducted important research on phosphorescence spectrum analysis, and the absorption of light. In 1903 Becquerel shared the Nobel Prize in physics with the French physicists Pierre Curie and Marie Curie for their work on radioactivity, a term Marie Curie coined. His works include Recherches sur la phosphorescence (Research on Phosphorescence, 1882-1897) and Decouverte des radiations invisibles émises par l'uranium (Discovery of the Invisible Radiation Emitted by Uranium, 1896-1897).

John W.S. Rayleigh

*"British physicist John W.S. Rayleigh won the 1904 Nobel Prize in physics. Rayleigh investigated the density of gases and shared credit for the discovery of argon, a gaseous element."*

# JOHN WILLIAM STRUTT RAYLEIGH

**Rayleigh, John William Strutt, 3rd Baron** (1842-1919), British mathematician, physicist, and Nobel laureate, known for his research in wave phenomena. Rayleigh was born in Lanford Grove, near Maldon, England, and educated at the University of Cambridge. He served as professor of experimental physics and director of the Cavendish Laboratory at Cambridge from 1879 to 1884 and as professor of natural philosophy at the Royal Institution, London, from 1887 to 1905. He became chancellor of Cambridge in 1908.

Rayleigh engaged in research into physical optics, light, color, and electricity and the dynamics of resonance and vibrations of gas and elastic solids. He was also responsible for the determination of electrical units of measurement. In 1894 he and the British chemist Sir William Ramsay discovered the inert element argon. For this discovery Rayleigh was awarded the 1904 Nobel Prize in physics. Ramsay won the Nobel Prize in chemistry the same year.

Philipp Lenard

*"German physicist Philipp Lenard won the 1905 Nobel Prize in physics. His research on cathode rays suggested atoms were made up primarily of empty space, a model upon which other physicists built."*

# PHILIPP EDUARD ANTON LENARD

**Lenard, Philipp Eduard Anton** (1862-1947), German physicist and Nobel Prize winner. Lenard was awarded the 1905 Nobel Prize in physics for his investigation of cathode rays (high-speed electrons produced in vacuum tubes), which contributed to the knowledge of the nature of atoms.

Born in Pressburg, Austria-Hungary (now Bratislava, Slovakia), his interest in physics drew him to the Universities of Berlin and Heidelberg in Germany. He spent two semesters in 1883 and 1884 at Berlin, and four semesters from 1884 to 1886 at Heidelberg, receiving a Ph.D. degree in physics from the University of Heidelberg in 1886. After academic posts in a number of German universities, he became a professor and the director of the physics laboratory at the University of Kiel, Germany, in 1898. Lenard returned to Heidelberg in 1907 as a professor and director of the Radiological Institute, where he remained until his retirement in 1931.

Lenard began working on cathode rays at the University of Bonn in 1892, where he was an assistant to German physicist Heinrich Hertz. Hertz had discovered that cathode rays could pass through a piece of thin metal foil in a vacuum tube, but not through the glass of the vacuum tube. Lenard, a talented experimentalist, modified vacuum tubes by adding an aluminum "Lenard window" so that the vacuum tube would still keep out air, but cathode rays could pass out of the tube and be studied more easily. He found that denser substances generally absorbed more cathode rays than did less dense substances, and that the rays penetrated substances more easily if they came from higher-voltage vacuum tubes. Lenard's research showed that cathode rays could pass through atoms, and this led him to conclude that the center of the atom was concentrated in a tiny fraction of the atomic volume. (He also suggested, incorrectly, that an atom's positive and negative charges occurred in tight pairs, which he termed *dynamids.*)

Other physicists subsequently developed more-precise models of the atom, but Lenard was the first to realize that the atom must have a structure, and the first to envision the atom as mostly empty space. Lenard's view was an important forerunner of British physicist Ernest Rutherford's 1911 model of the atom as a dense positive nucleus surrounded by orbiting negative electrons, which became the accepted model.

In addition to his Nobel Prize-winning work on cathode rays, Lenard's career included over 40 years of research in phosphorescence (the emission of light without combustion), and the first accepted explanation of the photoelectric effect (in which light falling on metal produces the emission of electrons from the metal surface). Both areas were important to the development of quantum theory.

Sir Joseph Thomson

*"British physicist Sir Joseph Thomson won the 1906 Nobel Prize in physics. Perhaps best remembered as the discoverer of the electron, he also conducted research into the conduction of electricity by gasses."*

# Sir Joseph John Thomson

**Thomson, Sir Joseph John** (1856-1940), British physicist and Nobel laureate. Thomson was born near Manchester, England, and educated at Owens College (now part of Victoria University of Manchester) and Trinity College, University of Cambridge. At Cambridge he taught mathematics and physics, served as Cavendish Professor of Experimental Physics, and was (1918-40) master of Trinity College. He was also president of the Royal Society (1915-20) and professor of natural philosophy at the Royal Institute of Great Britain (1905-18).

Thomson was awarded the 1906 Nobel Prize in physics for his work on the conduction of electricity through gases. He is considered the discoverer of the electron through his experiments on the stream of particles (electrons) emitted by cathode rays. A theorist as well as an experimenter, Thomson advanced in 1898 the "plum-pudding" theory of atomic structure, holding that negative electrons were like plums embedded in a pudding of positive matter. Thomson was knighted in 1908.

Albert A. Michelson

*"American physicist Albert A. Michelson won the 1907 Nobel Prize in physics. He developed extremely precise scientific instruments, one of which the interferometer, proved that the ether did not exist as a universal substance."*

# ALBERT ABRAHAM MICHELSON

**Michelson, Albert Abraham** (1852-1931), German-born American physicist, known for his famous experiment to measure the velocity of the earth through the ether, a substance that scientists believed filled the universe. This experiment helped prove that the ether does not exist. In 1907 he was awarded the Nobel Prize in physics for developing extremely precise instruments and conducting important investigations with them, becoming the first American citizen to earn a Nobel Prize.

Michelson was born in Strelno (now Strzelno, Poland), brought to the United States as a child, and educated at the United States Naval Academy and at the universities of Berlin, Heidelberg, and Paris. He was professor of physics at Clark University from 1889 to 1892, and from 1892 to 1929 was head of the department of physics at the University of Chicago. He determined the velocity of light with a high degree of accuracy, using instruments of his own design.

In 1887 Michelson invented the interferometer, which he used in the famous experiment, performed with the American chemist Edward Williams Morley. At that time, most scientists believed that light traveled in waves through the ether. They also believed that the earth traveled through the ether. The Michelson-Morley experiment showed that two beams of light sent in separate directions from the earth were reflected at the same speed. According to the theory of ether, the beams would have been reflected in waves of different speeds, in relation to the velocity of the earth. In this way, the experiment proved that the ether did not exist. The negative results of the experiment were also useful in the development of the theory of relativity. Michelson's major works include Velocity of Light (1902) and Studies in Optics (1927).

Gabriel Jonas Lippmann

*"French physicist Gabriel Lippmann won the 1908 Nobel Prize in physics. Lippman's invention of the first process for making permanent colour photographs earned him the award."*

# Gabriel Jonas Lippmann

**Lippmann, Gabriel Jonas** (1845-1921), French physicist and winner of the 1908 Nobel Prize in physics for his invention of the first process for making color photographs that did not fade quickly after development.

Born of French parents in Hollerich, Luxembourg, Lippmann was educated at home until age 13, when his family moved to Paris. He entered the École Normale Supérieure in 1868 and received a D. Sc. degree from the Sorbonne in 1875. In 1878 he joined the Faculty of Science in Paris, where he became professor of mathematical physics in 1883, then professor of experimental physics, and in 1886 director of the research laboratory, which was later transferred to the Sorbonne.

Lippmann's color photographic technique was based on interference, the combining of different light waves arriving simultaneously at the same point-the same phenomenon that causes color to appear in colorless substances such as soap bubbles. To receive the image, Lippmann used a glass plate coated on one side with a light-sensitive emulsion, a mixture of gelatin, grains of silver nitrate, and potassium bromide. In the camera, the emulsion side of the plate faced a plate holder coated with mercury, which acted as a mirror. When the camera lens was opened, light was reflected from the objects in the lens's field of view through the lens to the emulsion-coated plate and through the plate to the mirror; the various wavelengths of this light corresponded to the various colors of the objects in the field of view. The incoming light was then reflected back into the emulsion by the mirror. When the incoming light waves and the light waves reflected by the mirror met on the surface of the emulsion, they created interference patterns in the silver grains of the emulsion. These patterns were then fixed on the plate by chemical baths. When the plate dried, the interference patterns reflected light in various wavelengths corresponding to the original colors of the photographic objects. Lippmann's process was an important experimental milestone although it proved impractical in photography because exposure times were too lengthy, the image had to be viewed at a precise angle to a light source, and it could not be reproduced.

Lippmann was a multitalented researcher and made many other valuable scientific contributions, especially in electricity and optics. His textbook on *thermodynamics* (the physics of heat in relation to other energy forms) was the standard reference in France. He invented the capillary electrometer, which measured small differences in voltage and was used in early electrocardiographs. His research in piezoelectricity (the generation of electricity by compression or expansion of crystals such as quartz) and seismology (the measurement of earth tremors) furthered developments in those fields. Lippmann also invented the coelostat, a new astronomical tool that compensated for the earth's rotation and allowed a region of the sky to be photographed without apparent movement.

Karl Braun

*"German physicist and inventor Karl Braun won the Nobel Prize in physics in 1909. He made important contributions to wireless communication."*

# KARL FERDINAND BRAUN

**Braun, Karl Ferdinand** (1850-1918), German physicist, inventor, and Nobel Prize winner. Braun is best known for his invention of the first oscilloscope (an electronic instrument that displays changes in the voltage of an electric circuit) made out of a cathode-ray tube (CRT), but he also contributed much to the study of electricity and telegraphy, or wireless communication, through groundbreaking research and inventions. He shared the 1909 Nobel Prize in physics with Italian electrical engineer and inventor Guglielmo Marconi for their work on wireless communication.

Born in Fulda, Braun studied at the University of Marburg and received his doctorate from the University of Berlin in 1872, after a dissertation on the vibrations of elastic rods and strings. He began his career as a research assistant at the University of Würzburg, and later held positions at universities in Leipzig, Marburg an der Lahn, Karlsruhe, and Tübingen, where he founded the Physics Institute. From 1880 to 1883 Braun was at Strasbourg, France, and he returned there in 1895 to become professor of physics and director of the Strasbourg Physics Institute.

In 1874 Braun published some of the results of his research on mineral-metal sulfides. He found that these crystals conducted electric currents in only one direction. At the time, the information was important in electrical studies and in measuring electrical conductivity, but Braun's discovery found practical application in the early 20th century when it was employed in crystal radio receivers. The crystal rectifier allowed current to flow in only one direction and improved radio transmission.

Braun also created the first oscilloscope, then called a Braun tube, in 1897. The Braun tube was a valuable laboratory instrument, and modified versions are used in electronic testing and research today. The principle of the Braun tube, moving an electron beam by means of alternating voltage, is the principle on which all television tubes operate.

While Braun also made many contributions to pure science, he won honors for his fundamental modification of Marconi's wireless transmitting system. Braun tried to overcome the difficulty in increasing the range of a transmitter beyond 15 km (9 mi). He believed he could expand this range by increasing the power of the transmitter.

Studying the Marconi transmitter, which used a sparking apparatus to produce periodic waves that travel through the air, he learned that attempts to increase the power output by increasing the length of the spark gap eventually reached a limit at which the spark caused a decrease in output. Braun solved this dilemma by producing a sparkless antenna circuit. He magnetically coupled the power from the transmitter to the antenna circuit using a transformer effect instead of having the antenna directly in the power circuit.

Braun's circuit has been applied to all similar transmission, including radio, radar, and television. A patent was granted on this circuit in 1899. Braun also invented a transmitter that channeled the transmission of electric waves in one direction.

Braun gained notoriety outside the laboratory as well when he was called to the United States in 1914 to testify in litigation involving radio broadcasting. He was still in the country in 1917 when the United States entered World War I, and he was not allowed to return to Germany. He died in a Brooklyn, New York, hospital in 1918.

Guglielmo Marconi

*"Inventor of the radio-signaling system, Guglielmo Marconi was the first to send wireless signals across the ocean. Prior to his invention, there was no way to communicate over long distances without telegraph wires to carry electric signals. His equipment played a vital role in rescuing survivors of sea disasters such as the sinking of the Titanic. He won the Nobel Prize in physics in 1909 for his work in Wireless telegraphy."*

# Guglielmo Marconi

**Marconi, Guglielmo** (1874-1937), Italian electrical engineer and Nobel laureate, known as the inventor of the first practical radio-signaling system. He was born in Bologna and educated at the University of Bologna. As early as 1890 he became interested in wireless telegraphy, and by 1895 he had developed an apparatus with which he succeeded in sending signals to a point a few kilometers away by means of a directional antenna. After patenting his system in Great Britain, he formed (1897) Marconi's Wireless Telegraph Company, Ltd., in London. In 1899 he established communication across the English Channel between England and France, and in 1901 he communicated signals across the Atlantic Ocean between Poldhu, in Cornwall, England, and St. John's, in Newfoundland. His system was soon adopted by the British and Italian navies, and by 1907 had been so much improved that transatlantic wireless telegraph service was established for public use. Marconi was awarded honors by many countries and received, jointly with the German physicist Karl Ferdinand Braun, the 1909 Nobel Prize in physics for his work in wireless telegraphy. During World War I he was in charge of the Italian wireless service and developed shortwave transmission as a means of secret communication. In the remaining years of his life he experimented with shortwaves and microwaves.

Johannes Waals

*"Dutch physicist Johannes Waals won the 1910 Noble Prize in physics. Interested in thermodynamics, he developed a theory known as the van der Waals equation."*

# JOHANNES DIDERIK VAN DER WAALS

**Waals, Johannes Diderik van der** (1837-1923), Dutch physicist and Nobel laureate. He was born in Leiden and educated at Leiden University. From 1877 to 1907 he was professor of physics at the University of Amsterdam. van der Waals was interested primarily in thermodynamics; he developed a theory of corresponding states on the continuity of the liquid and gaseous states of matter expressed in the van der Waals equation. For these discoveries he was awarded the 1910 Nobel Prize in physics. He also studied the attractive forces holding the atoms of molecules together. These are called van der Waals forces, in his honor.

Wilhelm Wien

*"German physicist Wilhelm Wien won the 1911 Nobel Prize in physics. His discoveries in the field of radiation, including the laws that govern heat radiation, laid the foundation for the development of the quantum theory."*

# WILHELM WIEN

**Wien, Wilhelm** (1864-1928), German physicist and Nobel laureate, noted for his work on black-body radiation Heat. Wien was born in Gaffken and educated at the universities of Göttingen, Heidelberg, and Berlin. In 1890 he became an assistant to the German physicist Hermann Ludwig von Helmholtz at the Imperial Physical Technical Institute at Charlottenburg. At various times from 1900 until his death, Wien was professor of physics at the universities of Giessen, Würzburg, and Munich. He also visited (1913) the U.S. to lecture at Columbia University.

Wien developed a formula for determining the energy density associated with particular wavelengths for any given temperature of a radiating body. His contributions to the field of radiation laid the foundation for the development of the quantum theory. Wien conducted research in other fields, including optics and X rays. For his discovery of laws of heat radiation he was awarded the 1911 Nobel Prize in physics.

Nils Gustaf Dalen

*"Swedish engineer, inventor and acetylene gas expert Nils Gustaf Dalén won the Nobel Prize in physics in 1912. Dalén developed a solar valve that turned acetylene gas lights on and off automatically."*

# NILS GUSTAF DALEN

**Dalén, Nils Gustaf** (1869-1937), Swedish engineer, inventor, and Nobel Prize winner. Dalén received the 1912 Nobel Prize in physics for developing a solar valve, known as the sunvalve or Solventil, that turned acetylene-gas lights on and off automatically.

Born in Stenstorp, Dalén received a B.S. degree in mechanical engineering in 1896 from the Chalmers Institute in Göteborg, and attended the Federal Institute of Technology in Zürich, Switzerland, in 1896 and 1897. Returning to Sweden, he worked on improvements to hot-air turbines, compressors, and air pumps. In 1900 he cofounded the engineering firm of Dalén and Celsing. A year later, he became technical chief of the Swedish Carbide and Acetylene Company, later the Swedish Gas Accumulator Company. He was named managing director in 1909 and held that post until his death.

Dalén invented the Solventil in 1907. The Solventil had four metal rods suspended vertically inside a glass cylinder. Three of the rods were highly polished and surrounded the fourth, blackened and mounted in the center of the cylinder. Sunlight reflecting from the brighter rods warmed the black rod, which expanded and pressed a lever that closed a gas valve, extinguishing the flame. When sunlight faded, the central rod cooled and contracted, the valve opened, and the gas was ignited by a small, constantly lit flame called a bypass jet, or pilot light. By raising or lowering the four rods, the Solventil could be set to open at any degree of darkness.

Dalén became known as the "benefactor to sailors," since his bright-burning acetylene lights were used in lighthouses and marine buoys, as well as in railway signals. Widely used, the Solventil conserved gas and allowed lighthouses and beacons to operate unattended for months at a time. Even after many of the gaslit navigational lights had been converted to electricity, the Solventil was still used in some more remote locations. Another of Dalén's inventions, a device that provided thousands of rapid flashes from a single liter of gas, also benefited lighthouse operation.

Dalén made the highly explosive acetylene gas a safer fuel when he improved the gas accumulator, a storage system invented in France. Acetylene dissolved in acetone is nonexplosive, but as the solution is used, explosive gas accumulates in the space left above the liquid. Dalén developed a porous substance called aga, which he enclosed in a small steel container. He filled the container halfway with acetone and added acetylene under pressure, compressing the gas so that the container held 100 times as much acetylene as it would have under normal pressure. The aga remained in place as the solution was used, so no spaces developed. Dalén's gas accumulators could be handled without risk of detonation from jolting.

Though a 1912 laboratory explosion cost him his sight, Dalén produced other inventions, including an efficient cookstove and a device to change gas mantles (the finely woven metal mesh that glows, producing light, when heated) in unattended lighthouses.

Heike Kamerlingh Onnes

*"Dutch physicist Heike Kamerlingh Onnes won the 1913 Nobel Prize in physics. He conducted pioneering studies on how extremely low temperatures affect gases, metals and electrical conductivity."*

# HEIKE KAMERLINGH ONNES

**Kamerlingh Onnes, Heike** (1853-1926), Dutch physicist and Nobel laureate, born in Groningen, and educated at the University of Groningen. He was professor of physics at the Leiden University from 1882 until his death. Best known for his work in cryogenics, or the study of the production and effects of extremely low temperatures, he succeeded in liquefying helium for the first time in 1908. He studied the effects of extreme cold on a number of gases and metals and made significant discoveries relating to superconductivity, increased electrical conductivity at temperatures approaching absolute zero. He was awarded the 1913 Nobel Prize in physics.

Max von Laue

*"German physicist Max von Laue won the 1914 Nobel Prize in physics. His research revealed that X-rays cause distinct patterns of diffraction when passed through crystals."*

# MAX VON LAUE

**Laue, Max von** (1879-1960), German physicist and Nobel Prize winner. Laue received the 1914 Nobel Prize in physics for his discovery of X-ray crystallography (the study of patterns produced by the diffraction of X rays by crystal substances), which provided the means to determine the arrangement of atoms in some substances.

Born in Pfaffendorf, Laue started studying physics at the University of Strasbourg (which at the time was a German university; the city of Strasbourg is now part of France) in 1899, then continued his studies at the Universities of Göttingen, Munich, and Berlin, earning a Ph.D. degree in physics from the University of Berlin in 1903. He was an assistant to his mentor, physicist Max Planck, at the Institute for Physics in Berlin, and was affiliated with several German universities as well as the University of Zürich in Switzerland. In 1917 Laue became deputy director to American physicist Albert Einstein at the Institute for Physics. Later Laue added the duties of professor at the University of Berlin, where he taught until 1943, when he resigned in protest of the Nazi regime. In 1945, during World War II (1939-1945), Laue was taken prisoner by the Allies and was sent to England with other German scientists. He returned to Germany in 1946 and became head of the Max Planck Institute in Göttingen. In 1951, at age 71, he became director of the Fritz Haber Institute in Berlin, where he stayed until his retirement in 1958.

Optics and the wave theory of light interested Laue. After the 1895 discovery of X-rays by the German physicist Wilhelm Conrad Roentgen, scientists debated whether X-rays were particles or short electromagnetic waves. Physicists studied visible light by diffraction using a finely ruled grating. The spacings in a diffraction grating had to correspond with the wavelength of the light studied, and no artificial grating was fine enough for the estimated wavelength of X-rays, assuming they were electromagnetic waves.

At the University of Munich in 1912 Laue predicted that X-rays could be diffracted by a crystal acting as a natural diffraction grating. Faculty members Walter Friedrich and Paul Knipping undertook the investigation, using copper sulfate as the crystal, with immediate success. As Laue had predicted, dark spots on a photographic plate behind the crystal revealed not only the primary X-ray the researchers had directed at the crystal, but also the diffracted rays around it. Experiments with various crystals produced patterns, now known as Laue patterns, from which crystal structure could be interpreted.

Einstein called Laue's work one of the most beautiful discoveries in physics. It led to the development of X-ray spectroscopy, determination of X-ray wavelength, and the exploration of atomic structures of chemical elements. X-ray structural analysis became part of physics and chemistry, with practical applications in industry.

Laue later studied forces between atoms and worked on the thermodynamics of superconductivity. His books on the history of physics and Einstein's theory of relativity are widely read.

Sir Willliam Henry Bragg

*"British physicist Sir William Henry Bragg won the Nobel Prize in physics in 1915. He developed X-ray crystallography which was used to study crystal structure."*

# SIR WILLIAM HENRY BRAGG

**Bragg, Sir William Henry** (1862-1942), British physicist and Nobel laureate, born in Wigton, Cumberland, England, and educated at King William's College and at Trinity College, University of Cambridge. He was Cavendish Professor of Physics at the University of Leeds between 1909 and 1915. Bragg and his son Sir William Lawrence Bragg shared the 1915 Nobel Prize in physics-the only father-son team to win the Nobel Prize. Both also shared in other honors for their pioneer work in the study of crystal structure by measurement of X-ray diffraction. From 1915 to 1923, Sir William Henry, who was knighted in 1920, held the chair of physics at the University of London. During World War I, he headed a research group that invented the hydrophone, an instrument used for detection of submarines.

Sir William Lawernce Bragg

*"British physicist Sir William Lawrence Bragg won the Nobel Prize in physics in 1915. He measured X-ray diffraction to study crystal structure."*

# Sir William Lawrence Bragg

**Bragg, Sir William Lawrence** (1890-1971), Australian-born British physicist and Nobel Prize winner. Bragg shared the 1915 Nobel Prize in physics with his father, British physicist Sir William Henry Bragg, for their work in establishing X-ray crystallography, the study of crystal structures with X-rays.

Born in Adelaide, Australia, William Lawrence Bragg studied at Saint Peter's College in Adelaide and at the University of Adelaide, graduating in 1908. He enrolled at Trinity College, Cambridge, England, in 1909 to continue studying mathematics, but switched to physics at the suggestion of his father. Lawrence Bragg began research under the direction of British physicist Sir Joseph John Thomson in 1912. Bragg served in the British army during World War I (1914-1918), developing techniques to locate the enemy by the sound of their artillery fire. After the war, he held positions at Trinity College and then the University of Manchester. In 1937 Lawrence Bragg moved to the National Physical Laboratory as director, but soon accepted an invitation to Cambridge as the Cavendish Professor of Experimental Physics. He stayed at Cambridge until 1953, when he moved to the Royal Institution, London, as director of the Davy-Faraday Research Laboratory, a position once held by his father. He stayed at the Royal Institution until his retirement in 1966.

The work that brought the Braggs fame was based on the phenomenon of X-ray diffraction in crystals, discovered in 1912 by Max Theodor Felix von Laue. Although the wave nature of X-rays and the order of magnitude of their wavelength had been established, there were no methods developed to interpret the photographic interference pictures that two of von Laue's colleagues had produced by directing X-rays through crystals.

Lawrence Bragg and his father had begun discussing von Laue's findings in 1912, and worked together to treat them mathematically and simplify their interpretation. Lawrence Bragg discovered that certain planes in a crystal reflect X-rays, in accordance with the normal law of reflection (with equal angles of incidence and reflection). The distance between parallel planes of atoms determines the angle at which reflection can take place for a certain wavelength of the X-rays.

This relation, known as Bragg's law, permitted physicists to measure the wavelengths of the X-rays. For crystals of simple structure, like salt, physicists could calculate the distances between the planes of atoms from previous data and use the results to find the wavelength of the X-rays.

These discoveries served as the foundation for further research. The Braggs decided to build upon initial findings by using X-rays to study the structure of solids, especially crystals. With his father, Lawrence Bragg went on to determine the structure of increasingly complex compounds such as silicates, a family of crystalline minerals composed partly of silicon.

In his position at the Royal Institution, Bragg built upon his early work by sponsoring and conducting crystallographic research. He also devoted his energy and talents to popularizing and teaching science and its history. Bragg was knighted in 1941. He retired in 1966, but maintained an active interest in both crystallography and scientific popularization.

Charles Barkla

*"English physicist Charles Barkla won the Nobel Prize in physics in 1917. He investigated Roentgen radiation."*

# CHARLES GLOVER BARKLA

**Barkla, Charles Glover** (1877-1944), English physicist and Nobel laureate. Barkla devoted most of his career to investigations of X rays. For his study of radiation given off by substances when exposed to X rays, Barkla was awarded the 1917 Nobel Prize in physics.

Born in Widnes, England, Barkla studied mathematics and physics at University College in Liverpool, where he received a B.S. degree in 1898, an M.S. degree in 1899, and a Ph.D. degree in 1904. He taught at Liverpool until 1909, when he became Wheatstone Professor of Physics at King's College in London. From 1913 until his death, he held the chair of natural philosophy at the University of Edinburgh in Scotland.

When Barkla first noticed secondary radiation coming from substances exposed to X rays, he assumed that the secondary radiation was caused by scattering of the primary X rays. Since the intensity of the secondary radiation (composed mostly of electrons) increased with the density of the substance, Barkla concluded that the more massive the atoms and molecules of the substance, the more electrons they contained. This was the first suggestion of a link between the number of electrons in the atom of an element and that element's position in the periodic table Barkla later found that secondary radiation from elements with the heaviest atoms and molecules had two components. One was the X rays that scattered unchanged. A second, more penetrating type of radiation, called *characteristic radiation*, was produced by the element itself and had characteristics specific to the element. In further investigations, Barkla showed that two kinds of the second type of radiation (characteristic radiation) were produced by heavy elements. He named the more penetrating of these types K radiation and the less penetrating, or "soft," type, L radiation.

Characteristic radiation was essential to further research on the inner structure of atoms. It contributed to the work of English physicist Henry Gwyn-Jeffreys Moseley, who established the meaning of atomic number (the number of protons in an atom), and to Swedish physicist Karl Manne Siegbahn's analysis of the spectrum of X rays.

Barkla also showed that X rays were transverse waves like light, proving that they were electromagnetic radiation. Although he was an accomplished experimentalist and had a wide knowledge of physics, Barkla was a weak theorist. By 1916 he had rejected the quantum theory of German physicist Max Planck, German-born American physicist Albert Einstein, and Danish physicist Niels Bohr. Barkla's later investigations were aimed at finding the *J phenomenon* (radiation more penetrating then K radiation), but that research was not productive.

Max Planck

*"In a radical departure from classical ideas, theoretical physicist Max Planck proposed that energy travels in discrete packets called quanta. He won the Nobel Prize in physics in 1918. Prior to Planck's work with black body radiation, energy was thought to be continuous, but this theory left many phenomena unexplained. Planck realized that quantized energy could explain the behaviour of light. His revolutionary work laid the foundation for much of modern physics."*

# MAX KARL ERNST LUDWIG PLANCK

**Planck, Max Karl Ernst Ludwig** (1858-1947), German physicist and Nobel laureate, who was the originator of the quantum theory.

Planck was born in Kiel on April 23, 1858, and educated at the universities of Munich and Berlin. He was appointed professor of physics at the University of Kiel in 1885, and from 1889 until 1928 filled the same position at the University of Berlin. In 1900 Planck postulated that energy is radiated in small, discrete units, which he called quanta. Developing his quantum theory further, he discovered a universal constant of nature, which came to be known as Planck's constant. Planck's law states that the energy of each quantum is equal to the frequency of the radiation multiplied by the universal constant. His discoveries did not, however, supersede the theory that radiation from light or matter is emitted in waves. Physicists now believe that electromagnetic radiation combines the properties of both waves and particles. Planck's discoveries, which were later verified by other scientists, were the basis of an entirely new field of physics, known as quantum mechanics, and provided a foundation for research in such fields as atomic energy.

Planck received many honors for his work, notably the 1918 Nobel Prize in physics. In 1930 Planck was elected president of the Kaiser Wilhelm Society for the Advancement of Science, the leading association of German scientists, which was later renamed the Max Planck Society. He endangered himself by openly criticizing the Nazi regime that came to power in Germany in 1933 and was forced out of the society, but became president again after World War II. He died at Göttingen on October 4, 1947. Among his writings that have been translated into English are *Introduction to Theoretical Physics* (5 volumes, 1932-33) and *Philosophy of Physics* (1936).

Johannes Stark

*"German physicist Johannes Stark won the 1919 Noble Prize in physics. His research into the nature of electron wavelengths given off by light through a spectrum contributed to the understanding of the structure of the atom."*

# JOHANNES STARK

**Stark, Johannes** (1874-1957), German physicist and winner of the 1919 Nobel Prize in physics for his discovery of the nature of electron spectral lines (wavelengths given off by light through a spectrum). This discovery contributed to the understanding of the structure of the atom.

Stark was born in Schickenhof, Germany, and earned his Ph.D. degree in physics at the University of Munich in 1897. Though a talented experimental physicist, his abrasive personality caused conflicts that often led to career moves. He was an assistant and lecturer at the University of Göttingen, then a professor at the technical university in Hannover and the Universities of Griefswald, Wurzburg, and Aachen, all in Germany. He withdrew from academia in 1922 and used his Nobel Prize money to finance a porcelain factory in Germany, which later failed. Stark's Nazi connections helped him win powerful positions, first as president of the State Physical-Technical Institute in 1933 and then as president of the federally funded German Research Association in 1934.

Stark was interested in the behavior of ions (electrically charged atoms or groups of atoms) in electrical fields. In 1905 he used the Doppler effect (shift in the perception of sound or light waves caused by the movement of the source of the wave in relationship to the observer) to explain a shift in the frequencies of light waves radiated by swiftly moving ions. The shift, as explained by the Doppler effect, indicated that the electrons were moving at different speeds. In 1913 Stark maintained a concentrated electrical field in a ray tube and placed a photographic plate in the path of light rays that passed through a mixture of hydrogen and helium. The light radiated by each type of ion produces its own unique pattern of colored and dark lines called a spectrum. The spectra differ depending on the frequency of the light waves being emitted, and the frequency varies depending on the energy level of the light waves. When developed, Stark's photographic plate showed that the spectral line produced by the hydrogen ions had split into several lines. As electrons moved from one energy level to another, the frequency of the light they gave off changed, creating the split spectral line. Later research showed that Stark's work supported the quantum theory, which states that as electrons and other subatomic particles move from one orbit, or energy level, to another they give off or absorb energy in specific amounts, or quanta.

Stark had been at the forefront of the new physics and an advocate of the theories of Max Planck, Albert Einstein, and others, but in the years after 1913, he opposed their theories. During the Nazi regime, he used his authority to fight against modern theoretical physics, which he called "Jewish physics," and under Nazi law, to remove Jewish professors from universities. His actions contributed to the immigration of both Jewish and non-Jewish physicists to the United States, many of whom participated in the development of the atom bomb during World War II (1939-1945). In 1947 a German court classified Stark as a "leading Nazi" and sentenced him to four years in a labor camp.

Charles-Edouard Guillaume

*"Swiss physicist Charles-Édouard Guillaume won the Nobel Prize in physics in 1920. He discovered invar, a nickel-steel alloy."*

# CHARLES-EDOUARD GUILLAUME

**Guillaume, Charles-Édouard** (1861-1938), Swiss-born French physicist, metrologist (specialist in scientific measurement), and Nobel Prize winner. Guillaume received the 1920 Nobel Prize in physics for development of nickel-steel alloys (mixtures of the metals nickel and steel).

Guillaume was born in Fleurier, Switzerland. He earned a Ph.D. degree in physics at the Federal Institute of Technology in Zürich in 1882 and spent a brief time as an artillery officer in the Swiss Army. In 1883 he began a career with the newly established International Bureau of Weights and Measures in Sèvres, France, serving first as an assistant, then as associate director from 1902 to 1905, and finally as the bureau's director from 1915 to his retirement in 1936. He remained an honorary director of the bureau until his death.

Guillaume's work involved improvement of precision measuring instruments. Local metrology laboratories had problems with accurate measurement of length due to slight expansion and contraction of measuring rules caused by temperature changes. After spending years systematically testing alloys, Guillaume developed an economical nickel-steel alloy he called Invar. He chose that name because the length of Invar does not vary with temperature-that is, it is invariable. Guillaume's Invar was composed of 36 percent nickel, 0.4 percent manganese, 0.1 percent carbon, and 63.5 percent iron. Metrology laboratories measure length with Invar meters, and surveyors use Invar tapes or wires. One of the main uses of Invar today is in computer monitors.

Another measurement problem was caused by variable elasticity in metals; for instance, temperature changes caused stiffness and unreliable frequencies in steel tuning forks. Guillaume developed Elinvar, a nickel-chromium-iron alloy named for its invariable elasticity. Both Invar and Elinvar became important in accurate measurement of time; used for parts in clocks and watches, these alloys revolutionized the manufacture of timepieces. Another of Guillaume's alloys, Platinite, was an inexpensive substitute for platinum in lightbulb manufacture, an improvement that made electric lighting widely available.

Guillaume's first assignment at the Bureau of Weights and Measures was to study mercury thermometers, and his published research in precision thermometry became the standard text. In other research, he redetermined the exact volume of the liter at 1000.028 cubic centimeters. His many publications communicated progress in metrology to the international scientific community. He was an advocate of the metric system.

Albert Einstein

*"In 1905 German born American physicist Albert Einstein published his first paper outlining the theory of relativity. He won the Noble Prize in physics in 1921. It was ignored by most of the scientific community. In 1916 he published his second major paper on relativity, which altered mankind's fundamental concepts of space and time."*

# ALBERT EINSTEIN

**Einstein, Albert** (1879-1955), German-born American physicist and Nobel laureate, best known as the creator of the special and general theories of relativity and for his bold hypothesis concerning the particle nature of light. He is perhaps the most well-known scientist of the 20th century. Einstein was one of the fathers of the nuclear age. Einstein's famous equation, E equals m times c-squared (energy equals mass times the velocity of light squared), became a foundation stone in the development of nuclear energy. Einstein developed his theory through deep philosophical thought and complex mathematical reasoning. The great scientist was once reported to have said that only a dozen people in the world could understand this theory. However, Einstein always denied this report. Einstein was fond of classical music and played the violin. Einstein was never concerned about money. Publishers and editors from all parts of the world offered him huge sums for an autobiography. He never considered such offers. Finally, he did write his "Autobiographical Notes" because, as he put it, "it is a good thing to show those who are striving alongside of us, how one's own striving and searching appears, to one in retrospect." Those "Notes" were the only document even approaching an autobiography that Einstein ever wrote. He wrote them for a scholarly volume without asking for or getting any money in return. Einstein was married twice.

Niels Bohr

*"A Nobel Prize winner, Niels Bohr was known not only for his own theoretical work, but also as a mentor to younger physicists who themselves made important contributions to physical theory. He won the Nobel Prize in physics in 1922. As the director of the Institute for Theoretical Physics at the University of Copenhagen, Bohr gathered together some of the finest minds in the physics community, such as Werner Heisenberg and George Gawow. During the 1920's, the Institute was the source of many important works in quantum mechanics, and theoretical physics in general."*

# Niels Henrik David Bohr

**Bohr, Niels Henrik David** (1885-1962), Danish physicist and Nobel laureate, who made basic contributions to nuclear physics and the understanding of atomic structure.

Bohr was born in Copenhagen, the son of a physiology professor, and was educated at the University of Copenhagen, where he earned his doctorate in 1911. That same year he went to Cambridge, England, to study nuclear physics under British physicist Sir Joseph John Thomson, but he soon moved to Manchester to work with another British physicist, Ernest Rutherford.

Bohr's theory of atomic structure, for which he received the Nobel Prize in physics in 1922, was published in papers between 1913 and 1915. His work drew on Rutherford's nuclear model of the atom, in which the atom is seen as a compact nucleus surrounded by a swarm of much lighter electrons. Bohr's atomic model made use of quantum theory and the Planck constant (the ratio between quantum size and radiation frequency). The model posits that an atom emits electromagnetic radiation only when an electron in the atom jumps from one quantum level to another. This model contributed enormously to future developments of theoretical atomic physics.

In 1916 Bohr returned to the University of Copenhagen as a professor of physics, and in 1920 he was made director of the university's newly formed Institute for Theoretical Physics. There Bohr developed a theory relating quantum numbers to large systems that follow classical laws, and made other major contributions to theoretical physics. His work helped lead to the concept that electrons exist in shells and that the electrons in the outermost shell determine an atom's chemical properties. He also served as a visiting professor at many universities.

In 1939, recognizing the significance of the fission experiments of German scientists Otto Hahn and Fritz Strassmann, Bohr convinced physicists at a scientific conference in the United States of the importance of those experiments. He later demonstrated that uranium-235 is the particular isotope of uranium that undergoes nuclear fission. Bohr then returned to Denmark, where he was forced to remain after the German occupation of the country in 1940. Eventually, however, he was persuaded to escape to Sweden, under peril of his life and that of his family. From Sweden the Bohrs traveled to England and eventually to the United States, where Bohr joined in the effort to develop the first atomic bomb, working at Los Alamos, New Mexico, until the first bomb's detonation in 1945. He opposed complete secrecy of the project, however, and feared the consequences of this ominous new development. He desired international control.

In 1945 Bohr returned to the University of Copenhagen, where he immediately began working to develop peaceful uses for atomic energy. He organized the first Atoms for Peace Conference in Geneva, held in 1955, and two years later he received the first Atoms for Peace Award. In 1997 the International Union of Pure and Applied Chemistry announced that the chemical element with the atomic number 107 would be given the official name bohrium (Bh), in honor of Niels Bohr.

Robert Millikan

*"American physicist Robert Millikan won the 1923 Nobel Prize in physics. Millikan conducted a series of experiments that measured the electrical charge carried by an electron."*

# Robert Andrews Millikan

**Millikan, Robert Andrews** (1868-1953), American physicist, best known for his work in atomic physics. Millikan was born in Morrison, Illinois, and educated at Columbia University and the universities of Berlin and Göttingen. He joined the faculty of the University of Chicago in 1896, and in 1910 he became professor of physics there. He left the university in 1921 to become director of the Norman Bridge Laboratory of Physics of the California Institute of Technology. He was awarded the 1923 Nobel Prize in physics for his famous "oil-drop" experiments, which measured the charge on an electron and showed that the charge exists only as a whole number of units of that charge. His other contributions include important research on cosmic rays (which he named) and X rays, and the experimental determination of Planck's constant. He wrote technical studies and several books on the relationship between science and religion.

Karl Siegbahn

*"Swedish physicist Karl Siegbahn won the 1924 Nobel Prize in physics. He was awarded the prize for his work in the field of X-ray spectroscopy, a technique that measures the X-rays emitted from substances to learn more about their physical characteristics."*

# Karl Manne Georg Siegbahn

**Siegbahn, Kárl Manne Georg** (1886-1978), Swedish physicist and Nobel laureate. Siegbahn's research in X-rayspectroscopy and his development of instruments for precise measurement of X-ray wavelengths advanced the exploration of atomic structure. For his discoveries and research in the field of X-ray spectroscopy, Siegbahn was awarded the 1924 Nobel Prize in physics.

Born in Örebro, Sweden, Siegbahn attended the Physics Institute at the University of Lund, where he received a B.S. degree in 1908, an M.S. degree in 1910, and a Ph.D. degree in 1911. He became a lecturer at Lund in 1911 and from 1922 to 1937 was a professor of physics at the University of Uppsala. From 1937 until his retirement in 1964, he was Research Professor of Experimental Physics at the Royal Swedish Academy of Sciences in Stockholm and the first director of its Nobel Institute of Experimental Physics.

Siegbahn's early work was in electricity and magnetism but by 1914 his attention had turned to X-ray *Spectroscopy*, a technique for investigating the portion of the electromagnetic spectrum that contains X rays-electromagnetic radiation with shorter wavelengths and higher frequencies than those of visible light. X-ray *Spectroscopy* is based on the fact that each element, when bombarded by fast-moving electrons, emits X rays of a characteristic wavelength and frequency. For example, the X rays emitted by calcium will be different from those emitted by iron. The X-ray spectrometer is an instrument that measures and records the wavelengths of the emitted X-rays. Siegbahn, a talented instrument designer, improved the X-ray spectrometer in various ways so it detected and measured X-rays with more precision and allowed him to discover previously unknown series of X-rays.

Siegbahn's work contributed to the understanding of the atom and its structure and supported the prevailing model that electrons were arranged in spherical shells around the nucleus of an atom. His research yielded information about virtually all the elements from sodium to uranium and made possible the analysis of unknown substances. His 1923 *Spectroscopy* of X-Rays was a standard reference, his measurements of X-ray wavelengths were relied upon for their precision, and other physicists adopted his precise instrumentation. Siegbahn's research ultimately resulted in many current applications of X-ray *Spectroscopy* in such diverse fields as nuclear physics, chemistry, astrophysics, and medicine.

James Franck

*"German-American physicist and chemist James Franck won the Nobel Prize in physics in 1925. His experiments provided verification of quantum theory."*

## JAMES FRANCK

**Franck, James** (1882-1964), German American physicist, chemist, and Nobel laureate, born in Hamburg, and educated at the universities of Heidelberg and Berlin. He was professor of physics at several universities in Germany and the U.S. In collaboration with the German physicist Gustav Hertz, Franck conducted notable experiments on the effects produced by bombarding atoms with electrons. For that research, which provided experimental verification of the quantum theory, he shared the 1925 Nobel Prize in physics with Hertz. Franck is also noted for his important contributions to the study of photosynthesis.

Gustav Hertz

*"German physicist Gustav Hertz won the 1925 Nobel Prize in physics. He won the award for his research on the effect of the impact of electrons on atoms."*

# GUSTAV HERTZ

**Hertz, Gustav** (1887-1975), German physicist and Nobel laureate, born in Hamburg, and educated at the universities of Göttingen, Munich, and Berlin. In conjunction with the American physicist James Franck, Hertz studied the effect of the impact of electrons on atoms. As a result of these experiments, which were the first demonstration of the quantum theory of the German physicist Max Planck, Hertz and Franck were awarded the 1925 Nobel Prize in physics. Hertz served as professor of experimental physics at the University of Halle from 1925 to 1927 and at the Berlin Technische Hochschule from 1928 to 1935, when he became director of the Siemens Research Laboratory in Berlin. In 1945 he went to the USSR to continue his work in atomic research; he was awarded the Stalin Prize in 1951.

Jean Baptiste Perrin

*"French physicist Jean Baptiste Perrin won the 1926 Nobel Prize in physics. Perrin confirmed the existence of atoms and molecules, proving the discontinuous nature of matter."*

## JEAN BAPTISTE PERRIN

**Perrin, Jean Baptiste** (1870-1942), French Physicist, Nobel Laureate. Perrin proved the discontinuous (or atomic) nature of matter, discovered the sedimentation equilibrium (the mathematical basis for the settling of small particles in liquid), and thus confirmed the existence of molecules. For this work, he was awarded the Nobel Prize in physics in 1926. Born in Lille, France, he studied at the Ecole Normale Superieure, where he received his doctorate in 1897 with a dissertation on cathode rays and X rays. He was a lecturer in physical chemistry at the Sorbonne in Paris and a professor there from 1910 to 1940. During World War I, he worked on acoustic detection of submarines. He created the National Center for Scientific Research and, as France's Under-Secretary of State for Scientific research, popularized science for young people. After Germany's invasion of France during World War II, Perrin fled to New York where he gathered support for the French war effort and helped establish the Franco-Belgian School of Higher Studies.

Perrin's key contribution came from his investigations of Brownian motion. In 1827 botanist Robert Brown observed random movements of pollen grains suspended in water and found that particles in a liquid at "equilibrium" continuously moved, turned, rose, and fell. One suggested cause was bombardment by liquid molecules, but the existence of molecules was disputed at that time and for the rest of the century. In 1905, Einstein published a molecular theory of Brownian motion. Perrin provided the difficult experimental proof during 1908 to 1913 with his studies of sedimentation (sinking of suspended particles, caused by gravity). Perrin demonstrated that the displacement of uniform particles resulted in a certain vertical distribution. He used photography to make thousands of microscopic observations, counting particles at various depths in a drop of water. The number of particles, greatest at the bottom, decreased at each higher level. In undisturbed fluid, these concentrations remained constant, despite the force of gravity. Perrin also measured rotation of particles. His findings confirmed the molecular theory and allowed him to calculate molecular size and Avogadro's number (number of molecules in a mole, or the mass, in grams, equal to the molecular weight of the substance). His 1913 book, The Atoms, presented evidence for the discontinuous nature of matter and led to wide acceptance of the existence of atoms and molecules.

The Royal Society of London awarded him its 1896 Joule Prize and elected him a foreign member. He held many honorary degrees and belonged to the scientific academies of several countries, including the French Academy of Sciences, of which he was president in 1938. Perrin died at age 71 in New York, and his remains were placed in the Pantheon monument in Paris.

Arthur Holly Compton

*"American physicist Arthur Holly Compton won the Nobel Prize in physics in 1927. Compton confirmed that radiation has both particle and wave characteristics."*

# ARTHUR HOLLY COMPTON

**Compton, Arthur Holly** (1892-1962), American physicist and Nobel laureate whose studies of X rays led to his discovery in 1922 of the, so-called Compton effect. The Compton effect is the change in wavelength of high energy electromagnetic radiation when it scatters off electrons. The discovery of the Compton effect confirmed that electromagnetic radiation has both wave and particle properties, a central principle of quantum theory.

Compton was born in Wooster, Ohio, and educated at Wooster College and Princeton University. In 1923 he became professor of physics at the University of Chicago. While at the University of Chicago, Compton directed the laboratory where the first sustainable nuclear chain-reaction was performed. Compton also played a role in the development of the atomic bomb. From 1945 to 1953 Compton was chancellor of Washington University, and after 1954 he was professor of natural philosophy there. For his discovery of the Compton effect and for his investigation of cosmic rays and of the reflection, polarization, and spectra of X rays, he shared the 1927 Nobel Prize in physics with the British physicist Charles Wilson.

Charles T.R. Wilson

*"English physicist Charles T. R. Wilson won the 1927 Nobel Prize in physics. Inventor of the cloud chamber, a type of particle detector, he made visible the paths of sub-atomic particles, eventually enabling scientists to determine their mass and charge."*

# Charles Thomson Rees Wilson

**Wilson, Charles Thomson Rees** (1869-1959), Scottish physicist and Nobel laureate. Wilson invented the cloud chamber, which gave the first pictures of the paths of subatomic particles and became an essential tool in the fields of atomic and meteorological physics. For his discovery of the method of making the paths of electrically charged particles visible by the condensation of water vapor, Wilson shared the 1927 Nobel Prize in physics with American physicist Arthur Holly Compton.

Wilson was born in Glencorse in the former county of Midlothian, Scotland. He received a B.S. degree from Owens College (now the Victoria Institute of Manchester) in England in 1887 and a B.A. degree from the University of Cambridge in 1892. After teaching at Bradford Grammar School in Bradford, England, for four years Wilson returned to Cambridge in 1896 as a researcher and remained there, eventually as a professor, until he retired in 1936. He remained active in research, publishing his last paper at the age of 87.

Wilson first developed the cloud chamber in the late 1890s to study how water vapor and light interact. Physicists at that time thought that water droplets formed only around dust particles. Wilson established that water droplets can form around charged particles, or ions, in the absence of dust. He found that if he exposed the air in a chamber to X rays, many more droplets formed. He concluded the X rays give the air molecules an electrical charge, or ionize them.

As an ion moves through the cloud chamber, drops of water form around it. Because the ion moves very quickly, the string of drops of water in the air looks like a continuous path marking the movement of the ion. The path is especially apparent when a strong light is directed at the cloud chamber. If a magnetic field is applied to the cloud chamber, the ions will follow curved paths depending on the strength and nature of the charge, the mass of the ion, and the strength and direction of the magnetic field. The paths can be photographed for later analysis.

Wilson also intensely studied electrical conduction in air and applied his findings to devising ways to protect British airships from lightning and other discharges of electricity during World War I (1914-1918).

Sir Owen W. Richardson

*"British physicist Sir Owen W. Richardson won the 1928 Nobel Prize in physics. Richardson's theory of thermionics described how a heated metal emits electrically charged particles, leading to developments in telephony, radio, television, and X-ray technology."*

## SIR OWEN WILLANS RICHARDSON

**Richardson, Sir Owen Willans** (1879-1959), British physicist and Nobel Prize winner. Richardson received the 1928 Nobel Prize in physics for his particles from a heated (the emission of charged atomic particles from a heated surface), which is described in a law that bears his name, and which formed the basis for the development of telephone, radio, television, and X-ray technology.

Born in Dewsbury, Yorkshire, England, Richardson won a scholarship in 1897 to Trinity College, Cambridge, where he earned his B.S. degree in physics, chemistry, and botany in 1900 and continued his graduate studies in chemistry and physics. He was elected a fellow of Trinity College, London, in 1904. At the prestigious Cavendish Laboratory at Cambridge, Richardson began his work on thermionics, a phrase he coined, for which he would later become world renowned. In 1906 he moved to the United States to accept a position as head of the physics department at Princeton University, where he taught for seven years. Richardson's book, The Electron Theory of Matter (1914), based on his Princeton, lectures, became a classics text for physics scholars. In 1913, Richardson returned to England just before the start of World War I (1914-1918) to accept the Wheatstone Chair of Physics at King's College at the University of London. During the war, Richardson aided his country by working to improve the technology of military-communications systems. In 1924 he was appointed Yarrow Research Professor of the royal society, an honor that freed him from teaching duties to devote his full attention to research.

When Richardson began his thermionic research, many scientists believed that electron emissions were caused by reactions between hot filaments and residual gases, and thus were chemical in nature. Richardson, however, believed that negatively and positively charged radiation emanates directly from heated metal, and thus is physical in nature. Using platinum, and later tungsten, he proved that heating metal results in a loss of electrons from the metal's surface - the electrons evaporate in much the same way that water molecules evaporates from the surface of a lake. He formulated the equation that expresses that phenomenon, now called Richardson's law. Richardson also made a major contribution in the area of gyromagnetics (the study of the magnetic properties of a rotating electrical particles) for his theory, later refined by other scientists, that an object's magnetism caused by the movement of its electrons. He was knighted in 1939.

Louis de Broglie

*"French physicist Louis Victor de Broglie won the Nobel Prize in physics in 1929. He discovered the wave nature of electrons."*

# LOUIS-VICTOR DE BROGLIE

**Broglie, Louis-Victor de** (1892-1987), French physicist and Nobel laureate, who made major contributions to the theory of quantum mechanics with his studies of electromagnetic radiation. De Broglie was born in Dieppe and educated at the University of Paris. He tried to rationalize the dual nature of matter and energy, both of which he found to be composed of corpuscles and moving in waves. For his 1923 theory describing the wave nature of electrons, he was awarded the 1929 Nobel Prize in physics. He was elected to the Academy of Sciences (1933) and to the French Academy (1943). He was named professor of theoretical physics at the University of Paris (1928), permanent secretary of the Academy of Sciences (1942), and adviser to the French Atomic Energy Commission (1945). Several of his books have been translated into English, including *Matter and Light (1939), Revolution in Physics (1953), Current Interpretation of Wave Mechanics (1964),* and *Quantum, Space, and Time (1984).*

Chandrasekhara Venkata Raman

*"Indian physicist Sir Chandrasekhara Venkata Raman won the 1930 Nobel Prize in physics. He discovered that monochromatic light scatters into different frequencies when it passes through a transparent substance, a phenomenon now called the Raman effect."*

# Sir Chandrasekhara Venkata Raman

**Raman, Sir Chandrasekhara Venkata** (1888-1970), Indian physicist best known for his research on the molecular scattering of light. For his discovery of this effect, known as the Raman effect, he was awarded the 1930 Nobel Prize in physics.

Raman was born in Trichinopoly (now Tiruchirapalli) and educated at Presidency College in Madras (now Chennai). He was professor of physics at the University of Calcutta (now Kolkata) from 1917 to 1933 and in the latter year was appointed head of the department of physics of the Indian Institute of Science in Bangalore. In 1947 he became director of the Raman Research Institute, also in Bangalore. He was knighted in 1929 and was named president of the Indian Academy of Sciences in 1934. Raman also studied the physical nature of musical sounds and the mechanics of musical instruments. He wrote *Molecular Diffraction of Light (1922)* and *The New Physics; Talks on Aspects of Science (1951).*

Werner Heisenberg

*"German physicist Werner Heisenberg is best known for his discovery of the uncertainty principle. This principle states that it is impossible to measure the exact position and the momentum of a given body at the same time. Heisenberg was a leader in the emerging field of quantum theory and his work earned him the 1932 Nobel Prize in physics."*

# WERNER HEISENBERG

**Heisenberg, Werner** (1901-1976), German physicist and Nobel Prize winner, who played a large part in the development of quantum mechanics. Quantum mechanics describes matter in terms of both particles and waves. One of Heisenberg's best known contributions to quantum theory is the uncertainty principle, which states that scientists cannot measure both the position and velocity (direction and speed) of a particle at the same time-they can only know one of the values with certainty.

Heisenberg was born in Würzburg, Germany. His family moved to Munich in 1910, where Heisenberg received his early education. In the summer of 1920 he graduated from a Munich gymnasium (the German equivalent to a United States high school) and entered the University of Munich. During his first two years of studies, he published four physics research papers, making Heisenberg-at age 20-one of the top contributors to theoretical physics research. Heisenberg finished his undergraduate and graduate work in three years, and in 1923 presented his doctoral dissertation on turbulence in streams of fluid.

In his early career Heisenberg was at the forefront of dramatic changes taking place in the field of quantum mechanics. He studied with three leading quantum theorists at three major centers of quantum research of that time: German physicist Arnold Sommerfeld at the University of Munich; German physicist Max Born at the University of Göttingen in 1923; and, from 1924 to 1927, Danish physicist Niels Bohr at the Institute for Theoretical Physics in Copenhagen.

Heisenberg developed the first version of quantum mechanics, called matrix mechanics, in 1925. His version explained the motion of electrons (tiny negatively charged particles) in an atom in purely mathematical terms. His equations showed why electrons behave the way they do, which scientists had been unable to explain before. Heisenberg realized that the laws of classical physics did not govern events on the quantum level. For example, electrons do not follow the laws of classical physics and orbit the nucleus of an atom in a defined path, as planets orbit the Sun.

Heisenberg's matrix mechanics predicted that molecular hydrogen (hydrogen made up of pairs of atoms, sharing their electrons to form molecules) should exist in two distinct forms, called orthohydrogen and parahydrogen. These two forms result from a property of atoms called spin, a kind of angular momentum. In 1925 Heisenberg predicted that the *spin* of the two hydrogen atoms was the same in parahydrogen, and opposite each other in orthohydrogen. Other scientists soon confirmed his prediction experimentally. Heisenberg won the 1932 Nobel Prize in physics for his development of quantum mechanics and his prediction of the two types of molecular hydrogen.

With the development of matrix mechanics, Heisenberg became one of the founders of quantum mechanics. At about the same time Heisenberg developed matrix mechanics, Austrian physicist Erwin Schrödinger developed a way to describe particles in terms of the probability that any of their characteristics would be a certain value. Schrödinger later showed that both his approach and Heisenberg's approach yielded the same result.

In 1927 Heisenberg became a professor of theoretical physics at the University of Leipzig. That year he published a paper explaining the uncertainty principle, which stemmed from his matrix mechanics. Using calculations that explain the motion of particles, he showed that it is impossible to know accurately both the velocity and position of a particle at the same time. The more accurately scientists measure one quantity, the more uncertainty exists in the measurement of the other. The consequence of the uncertainty principle is that a description in quantum mechanics is limited to a statement of the relative probability of a value rather than exact numbers.

In 1941 Heisenberg became a professor at the University of Berlin and director of the Kaiser Wilhelm Institute for Physics. During World War II (1939-1945) he chose to remain in Nazi Germany while many of his colleagues fled the country. He was the leader of Germany's atomic research team, despite his opposition to Nazi policies. He worked with Otto Hahn, one of the discoverers of nuclear fission, but the German team failed to develop nuclear weapons.

At the end of the war the United States arrested Heisenberg for his role in the German weapons program and detained him for nine months in England. Following his return to Germany in 1946, he became professor of physics and the director of the Max Planck Institute for Physics and Astrophysics (the former Kaiser Wilhelm Institute) in Göttingen. The institute moved to Munich in 1958, and Heisenberg moved with it, continuing as its director until his death.

Paul Dirac

*"Nobel Laureate Paul Dirac was an important contributor to modern quantum theory in physics. He won the Nobel Prize in 1933. By including relativistic considerations in the wave theory of electrons, Dirac predicted the magnetic properties of electrons and accounted for the electrons's spin"*

# Paul Adrien Maurice Dirac

**Dirac, Paul Adrien Maurice** (1902-1984), British theoretical physicist and Nobel laureate, renowned for his prediction of the existence of the positron, or antielectron, and for his research in quantum theory.

Dirac was born in Bristol, England, and educated at the universities of Bristol and Cambridge. His quantum theory of electron motion led him in 1928 to postulate the existence of a particle identical to the electron in every aspect but charge, the electron having a negative charge and this hypothetical particle a positive one. Dirac's theory was confirmed in 1932 when the American physicist Carl Anderson discovered the positron. In 1933 Dirac shared the Nobel Prize in physics with the Austrian physicist Erwin Schrödinger, and in 1939 he was made a fellow of the Royal Society. He was a professor of mathematics at Cambridge from 1932 to 1968, a professor of physics at Florida State University from 1971 until his death, and a member of the Institute for Advanced Study periodically between 1934 and 1959. Dirac's writings include *Principles of Quantum Mechanics* (1930).

Erwin Schrodinger

*"A pioneer in the area of quantum theory, Erwin Schrodinger is best known for his mathematical theory describing the wave mechanics of electrons. He and British physicist Paul Dirac shared the 1933 Nobel Prize for their contributions to the understanding of quantum mechanics."*

# ERWIN SCHRÖDINGER

**Schrödinger, Erwin** (1887-1961), Austrian physicist and Nobel Laureate. Schrödinger formulated the theory of wave mechanics, which describes the behavior of the tiny particles that make up matter in terms of waves. Schrödinger formulated the Schrödinger wave equation to describe the behavior of electrons (tiny, negatively charged particles) in atoms. For this achievement, he was awarded the 1933 Nobel Prize in physics with British physicist Paul Dirac and German physicist Werner Heisenberg, who also made important advances in the theory of atomic structure. *See also* Quantum Theory; Atom.

Schrödinger was born in Vienna, Austria. His father was an oilcloth manufacturer who had studied chemistry, and his mother was the daughter of a chemistry professor. He attended an elementary school in Innsbruck for a few weeks, but Schrödinger received most of his early education from a private tutor. In 1898 he entered the Gymnasium in Vienna, where he studied mathematics, physics, and ancient languages. He then attended the University of Vienna from 1906 to 1910, specializing in physics. Schrödinger obtained his doctoral degree in physics in 1910. After a year in military training, he returned to the university to teach a first-year physics laboratory class. His early research ranged over many topics in experimental and theoretical physics.

During World War I (1914-1918) Schrödinger served as an artillery officer and then returned to his previous post at Vienna. Conditions were difficult in Austria after the war, and in 1920 Schrödinger decided to go to Germany. After a series of short-lived posts at the University of Jena, Stuttgart University, and the

University of Breslau (now Wroc³aw, Poland) in 1920 and 1921, he became a professor of physics at the University of Zürich in Switzerland in 1921.

Schrödinger's most important work was done at Zürich, and his work received much attention. He succeeded German physicist Max Planck as professor of theoretical physics at the University of Berlin in 1927. Schrödinger remained there until the rise of the National Socialist (Nazi) movement in 1933, when he went to the University of Oxford in England. There he became a fellow of Magdalen College. Homesick, he returned to Austria in 1936 to take up a post at Graz University, but the Nazi takeover of Austria in 1938 placed Schrödinger in danger. Schrödinger was not Jewish, but his opposition to Nazi policies made him a potential target. The prime minister of Ireland, Eamon De Valera, helped Schrödinger get out of Austria. De Valera's help also led to an appointment to a post at the Institute for Advanced Studies in Dublin in 1939. Schrödinger continued work in theoretical physics in Dublin until 1956, when he returned to Austria to a chair at the University of Vienna. He stayed at the University of Vienna until his death.

Schrödinger's great discovery of wave mechanics originated with the work of French physicist Louis de Broglie. In 1923 de Broglie used ideas from German American physicist Albert Eintstein's special theory of relativity to show that an electron, or any other particle, has a wave associated with it De Broglie's work resulted in the equation ë = h/p, where ë is the wavelength of the associated wave, h is a number called Planck's constant, and p is the momentum of the particle. Physicists immediately deduced that if particles (particularly electrons) have waves, then a particular type of partial differential equation known as a wave equation should be able to describe their behavior. These ideas were taken up by both de Broglie and Schrödinger, and in 1926 each published the same wave equation. Unfortunately, while the equation is true, it was of very little help in explaining the behavior of particles.

Later the same year Schrödinger used a new approach. He studied the mathematics of partial differential equations and the Hamiltonian function, a powerful idea in mechanics developed by British mathematician Sir William Rowan Hamilton in the mid-

1800s. Schrödinger formulated an equation in terms of the energy of the electron and the energy of the electric field in which it was situated. Partial differential equations have many solutions, but solutions to Schrödinger's equation had to meet strict conditions to be useful in describing the electron. Among other things, they had to be finite and possess only one value. These solutions were associated with special values of the electron's energy level, known as *proper values or eigenvalues*.

Schrödinger solved the equation for the hydrogen atom, $V = -e^2/r$, in which V is the energy of the electric field surrounding the electron, e is the electron's charge, and r is its distance from the atom's nucleus. He found that the eigenvalues of the electron's energy corresponded with those of the energy levels given in the older theory of Danish physicist Niels Bohr. Bohr's theory of the atom described electrons orbiting atoms in strict circular orbits at particular distances that corresponded to specific levels of energy. In the hydrogen atom (which consists of one electron and one proton), the wave function Schrödinger derived instead describes where physicists are most likely to find the electron. The electron is most likely to be where Bohr predicted it to be, but it does not follow a strictly circular orbit. The electron is described by the more complicated notion of an *orbital* - a region in space where the electron has varying degrees of probability of being found.

Schrödinger's wave equation can describe atoms other than hydrogen as well as molecules and ions (atoms or molecules with electric charge), but such cases are very difficult to solve. In a few such cases physicists have found approximate solutions, usually with a computer carrying out the numerical work.

Schrödinger's mathematical description of electron waves found immediate acceptance. The mathematical description matched what scientists had learned about electrons by observing them and their effects. In 1925, a year before Schrödinger published his results, German-British physicist Max Born and German physicist Werner Heisenberg developed a mathematical system called matrix mechanics. Matrix mechanics also succeeded in describing the structure of the atom, but it was totally theoretical. It gave no picture of the atom that physicists could verify observationally. Schrödinger's vindication of de Broglie's idea of electron waves immediately overturned matrix mechanics, though later physicists showed that wave mechanics is equivalent to matrix mechanics.

During his later years Schrödinger became increasingly worried by the uncertain nature of quantum mechanics, of which wave mechanics is a part. Schrödinger believed he had produced a defining description of the atom in the same way that the three laws of English physicist Isaac Newton defined classical mechanics and the way that the equations of British physicist James Clerk Maxwell described electrodynamics. Instead, each new discovery about the structure of the atom only made atomic structure more complicated. Much of Schrödinger's later work was concerned with philosophy, particularly as applied to physics and the atom.

Sir James Chadwick

*"Bristish physicist Sir James Chadwick won the Nobel Prize in physics in 1935. Chadwick's discovery of the neutron led to the development of nuclear fission and the atomic bomb."*

## SIR JAMES CHADWICK

**Chadwick, Sir James** (1891-1974), British physicist and Nobel laureate, who is best known for his discovery in 1932 of one of the fundamental particles of matter, the neutron, a discovery that led directly to nuclear fission and the atomic bomb. He was born in Manchester and educated there at Victoria University. In 1909 he began working under the British physicist Lord Ernest Rutherford. At the end of World War I he went to the University of Cambridge with Rutherford, with whom he continued a fruitful collaboration until 1935. In that year Chadwick became professor at the University of Liverpool. From 1948 to 1958 he was master, and from 1959 a fellow, of Gonville and Caius College, Cambridge.

Chadwick was one of the first in Britain to stress the possibility of the development of an atomic bomb and was the chief scientist associated with the British atomic bomb effort. He spent much of his time from 1943 to 1945 in the United States, principally at the Los Alamos Scientific Laboratory at Los Alamos, New Mexico. A fellow of the Royal Society, Chadwick received the 1935 Nobel Prize in physics and was knighted in 1945.

Carl David Anderson

*"American physicist Carl David Anderson won the Nobel Prize in physics in 1936. Anderson discovered the positron, a fundmental subatomic particle."*

## Carl David Anderson

**Anderson, Carl David** (1905-1991), American physicist and Nobel laureate. Anderson was born in New York City and educated at the California Institute of Tech--nology, where he attained full professorial rank in 1939. In 1932 he discovered the positron, or positive electron, one of the fundamental subatomic particles. For this achievement he was awarded, with Victor Franz Hess, the 1936 Nobel Prize in physics. In 1936 Anderson also confirmed experimentally the existence of the elementary nuclear particle called the meson, which had been predicted in 1935 by the Japanese physicist Yukawa Hideki.

Victor Hess

*"Austrian-American physicist Victor Hess won the 1936 Nobel Prize in physics. He won the award for his research into cosmic rays."*

## VICTOR FRANZ HESS

**Hess, Victor Franz** (1883-1964), Austrian-American physicist and Nobel laureate, one of the earliest workers in the field of cosmic rays. He taught in both Austria and the United States. As early as 1911 he measured cosmic-ray activity at altitudes as great as 9000 m (30,000 ft). He shared the 1936 Nobel Prize in physics with the American physicist Carl David Anderson. Hess wrote *Conductivity of the Atmosphere (1928)* and *Cosmic Rays and Their Biological Effects (1949)*.

Part
III

*1937 Clinton J. Davisson (United States)*
*Sir George P. Thomson (United Kingdom)*
*1938 Enrico Fermi (Italy)*
*1939 Ernest O. Lawrence (United States)*
*1940 Not awarded*
*1941 Not awarded*
*1942 Not awarded*
*1943 Otto Stern (United States)*
*1944 Isidor I. Rabi (United States)*
*1945 Wolfgang Pauli (United States)*
*1946 Percy W. Bridgman (United States)*
*1947 Sir Edward V. Appleton (United Kingdom)*
*1948 Patrick M.S. Blackett (United Kingdom)*
*1949 Hideki Yukawa (Japan)*
*1950 Cecil F. Powell (United Kingdom)*
*1951 Sir John D. Cockcroft (United Kingdom)*
*Ernest T.S. Walton (Ireland)*
*1952 Felix Bloch (United States)*
*Edward M. Purcell (United States)*
*1953 Frits Zernike (Netherlands)*
*1954 Max Born (West Germany)*
*Walther Bothe (East Germany)*
*1955 Polykarp Kusch (United States)*
*Willis Eugene Lamb, Jr. (United States)*
*1956 William B. Shockley (United States)*
*Walter H. Brattain (United States)*
*John Bardeen (United States)*
*1957 Tsung-Dao Lee (China-United States)*
*Chen Ning Yang (United States)*
*1958 Pavel A. Cherenkov (Union of Soviet Socialist Republics)*
*Il'ja M. Frank (Union of Soviet Socialist Republics)*
*Igor Y. Tamm (Union of Soviet Socialist Republics)*
*1959 Owen Chamberlain (United States)*
*Emilio G. Segré (United States)*
*1960 Donald A. Glaser (United States)*
*1961 Robert Hofstadter (United States)*
*Rudolf L. Mössbauer (West Germany)*
*1962 Lev. D. Landau (Union of Soviet Socialist Republics)*
*1963 Eugene P. Wigner (United States)*
*Maria G. Mayer (United States)*
*Johannes Hans D. Jensen (West Germany)*
*1964 Nikolay G. Basov (Union of Soviet Socialist Republics)*
*Aleksandr Mikhailovich Prokhorov (Union of Soviet Socialist Republics)*
*Charles H. Townes (United States)*

Clinton Joseph Davisson

*"American physicist Clinton Joseph Davisson won the Nobel Prize in physics in 1937. Davisson found that electrons could be diffracted by crystals just like X-rays. He also proved they can show wavelike structure."*

# Clinton Joseph Davisson

**Davisson, Clinton Joseph** (1881-1958), American physicist and Nobel laureate. Davisson made major contributions to understanding the diffraction of electrons by crystals and shared the 1937 Nobel Prize in physics with British physicist Sir George Paget Thomson for their independent proof of the wave properties of electrons.

Born in Bloomington, Illinois, Davisson entered the University of Chicago in 1902. He left to teach physics at Purdue University in 1903 and 1904. In 1905 he became a physics instructor and research assistant at Princeton University. Returning to the University of Chicago for summer sessions, Davisson finally completed his B.S. degree in physics in 1908. After earning a Ph.D. degree in physics from Princeton University in 1911, Davisson accepted an appointment as an assistant professor of physics at the Carnegie Institute of Technology.

In 1917, during World War I (1914-1918), Davisson took leave from the Carnegie Institute to work on a military telecommunications project at the engineering department of the Western Electric Company Laboratories, which later became the Bell Telephone Laboratories. After the war, he decided to remain at the laboratories, which guaranteed him freedom to do full-time research, an uncommon opportunity in commercial laboratories at that time. Davisson retired from Bell in 1946 and accepted a position as a visiting professor of physics at the University of Virginia in Charlottesville, where he remained until his retirement in 1954.

At Western Electric, Davisson pursued his interest in thermionics, the emission of electrically charged particles, or *ions*, from conducting materials such as metals heated to high temperatures. His experiments in thermionics revised classic theory on conduction and the thermal energy of electrons.

In 1925 an accidental explosion of a bottle of *liquid air* (air that has been cooled and compressed until it becomes a liquid; see Matter, States of) in Davisson's laboratory put him on the path to a great discovery. The explosion allowed a piece of nickel Davisson was using as a target for high-speed electrons to become *oxidized*-that is, for the outer nickel atoms to bond with oxygen atoms. After cleaning the target by means of prolonged heating, Davisson found that although it had originally been composed of many small crystals, it had now formed several large crystals. He continued with the experiment but found that the electrons that now bounced off the target were bouncing off at completely different angles than before. Davisson attributed the change only to the different way the target was reflecting the electrons, but a 1926 meeting of physicists convinced him to search for a further cause. There he first learned in detail about the hypothesis of French physicist Louis de Broglie that any material particle could behave like a wave. Joining forces in 1926 with American physicist Lester Halbert Germer, Davisson searched for evidence of interference-that is, the effects resulting from adding waves together-between the electrons as they bounced off the target; an interference pattern would be present only if the electrons were behaving like waves-purely material particles cannot produce interference patterns.

In 1927 Davisson showed that, like electromagnetic waves, electrons can produce interference patterns and be diffracted by crystals. His experiments proved that electrons can show wavelike behavior and helped show that all matter can show wavelike behavior since electrons are one of the subatomic particles that make up all matter. For this discovery, Davisson was awarded the Comstock Prize of the National Academy of Sciences in 1928.

In the 1930s Davisson's research continued to focus on electron waves, especially in their application to crystal physics and electron microscopy and helped to develop a technique for electron focusing. Davisson's later research focused on the theory and application of electron optics, the theory of electronic devices and solid-state physics, also known as condensed-matter physics.

Sir George Thomson

*"British physicist Sir George Thomson won the 1937 Nobel Prize in physics. He discovered the diffraction of electrons by crystals, proving their wave pattern."*

# Sir George Paget Thomson

**Thomson, Sir George Paget** (1892-1975), British physicist and Nobel laureate, son of Sir Joseph Thomson. Thomson was born in Cambridge, England, and educated at Trinity College, University of Cambridge. He taught physics at Cambridge (1919-22), at the University of Aberdeen, Scotland (1922-30), and at the Imperial College of Science and Technology, London (1930-52). He was master of Corpus Christi College, University of Cambridge, from 1952 to 1962. He shared the 1937 Nobel Prize in physics with the America physicist Clinton Davisson for their independent proof of the wave properties of electrons. Thomson, also noted for his work in aerodynamics and nuclear energy, was knighted in 1943.

Enrico Fermi

*"Famous for producing the first controlled nuclear reaction in 1942, physicist Enrico Fermi worked as a consultant for the Manhattan Project during world War II, helping to design the atomic bomb. He won a Nobel Prize in 1938 for his work on artificial radioactivity. He inspired many students and continues to be honoured through various awards and institutions that were established in his name such as the Fermi National Accelerator Laboratory in Batavia, Illinois."*

# Enrico Fermi

**Fermi, Enrico** (1901-1954), Italian-born American physicist and Nobel Prize winner, who made important contributions to both theoretical and experimental physics. Fermi's most well-known contribution was the demonstration of the first controlled atomic fission reaction. Atomic fission occurs when an atom splits apart. Fermi was the first scientist to split an atom, although he misinterpreted his results for several years. He also had an important role in the development of fission for use as an energy source and as a weapon. He won the 1938 Nobel Prize in physics for his work in bombarding atoms with neutrons, subatomic particles with no electric charge. Initially, Fermi believed that this process created new chemical elements heavier than uranium, but other scientists showed that he actually split atoms to create fission reactions.

Ernest O. Lawrence

*"American physicist Ernest O. Lawrence won the 1939 Nobel Prize in physics. He invented the cyclotron, a device used to accelerate nuclear particles."*

# Ernest Orlando Lawrence

**Lawrence, Ernest Orlando** (1901-1958), American physicist and Nobel laureate, best known for his invention and development of the cyclotron, a device to accelerate nuclear particles and used in the discovery of the transuranium elements. Born in Canton, South Dakota, Lawrence was educated at the universities of South Dakota, Minnesota, and Chicago and at Yale University. He was appointed assistant professor of physics at the University of California in 1927 and full professor in 1930. The following year he founded the university radiation laboratory in Berkeley, becoming its director in 1936. He was awarded the 1939 Nobel Prize in physics and the Enrico Fermi Award in 1957.

Otto Stern

*"German-American physicist Otto Stern won the 1943 Nobel Prize in physics. He developed a method for studying the magnetism of atoms, atomic nuclei and protons."*

# OTTO STERN

**Stern, Otto** (1888-1969), German American physicist and Nobel laureate, known for his studies of molecular beams. Stern was born in Sorau, Germany (now ˉary, Poland), and educated at the University of Breslau. He taught at Technische Hochschule in Zürich and at the universities of Frankfurt and Hamburg. In 1933 he moved to the U.S., accepting the position of research professor of physics at the Carnegie Institute of Technology (now Carnegie Meilon University) in Pittsburgh, Pennsylvania. Stern developed a method of studying the magnetic moments of atoms, atomic nuclei, and protons by directing beams of atoms or molecules through magnetic fields. For this achievement he was awarded the 1943 Nobel Prize in physics.

Isidor Rabi

*"American physicist Isidor Rabi won the 1944 Nobel Prize in physics. He developed a method to measure the hyperfine structure of atomic spectra."*

## ISIDOR ISAAC RABI

**Rabi, Isidor Isaac** (1898-1988), American physicist and Nobel laureate, known for his development of the magnetic resonance method for the study of the hyperfine structure of atomic spectra. Rabi was born in Rymanów, Austria (now in Poland), and received his Ph.D. degree from Columbia University in 1927 for a dissertation on the magnetic properties of crystals. In 1930 Rabi began research on the magnetic properties of atomic nuclei in an effort to ascertain the nature of the force binding the protons in the nuclei. He was awarded the 1944 Nobel Prize in physics for his invention of a method to measure the hyperfine transition levels of atoms by using known radio frequencies to cause transitions between these energy levels. Rabi worked on radar development during World War II and was one of the leading scientists on the project that developed the atomic bomb.

Wolfgang Pauli

*"Wolfgang Pauli won the 1945 Nobel Prize in physics for his discovery of the exclusion principle, also called the Pauli principle, which states that no two electrons in an atom can have identical sets of quantum numbers. These numbers define an electron's energy, and the exclusion principle allowed scientists to describe the arrangement and behaviour of electrons in the chemical elements."*

# Wolfgang Pauli

**Pauli, Wolfgang** (1900-1958), Austrian-born Swiss physicist and Nobel laureate, known for his definition of the exclusion principle in quantum mechanics. He was born in Vienna, and educated at the University of Munich. He taught physics at the Universities of Göttingen (1921-1922), Copenhagen (1922-1923), and Hamburg (1923-1928) and was professor of theoretical physics at the Federal Institute of Technology in Zürich from 1928 until his death in 1958. He also served as visiting professor at the Institute for Advanced Study at Princeton, New Jersey (1935-1936, 1940-1945, 1949-1950, and 1954).

In 1925 Pauli defined the exclusion principle, also called the Pauli exclusion principle, which states that no two electrons can occupy the same quantum state simultaneously in an atom. Four quantum numbers-principle, angular momentum, magnetic, and spin quantum numbers-define the quantum state of an electron. Pauli's exclusion principle states that no two electrons in the same atom can share all four quantum numbers. Pauli's hypothesis in 1930 of the existence of the neutrino, a subatomic particle, was a fundamental contribution to the development of meson theory. He was awarded the 1945 Nobel Prize in physics.

Percy Bridgman

*"American physicist Percy Bridgman won the Nobel Prize in physics in 1946. Bridgman studied the properties of materials under high pressure and made it possible to synthesize diamonds."*

# Percy. Williams Bridgman

**Bridgman, Percy Williams** (1882-1961), American physicist and Nobel laureate, who was noted for his study of the behavior of materials at high pressure. Bridgman was born in Cambridge, Massachusetts, and educated at Harvard University. He joined the physics department of Harvard in 1910 and was appointed a full professor in 1919.

In his study of high-pressure phenomena, Bridgman was often forced to develop his own experimental equipment. Eventually he was able to create pressures as high as 400,000 atmospheres. Bridgman did experiments that explored the mechanical and thermodynamic properties of materials at high pressure. In addition to the scientific discoveries he made using his equipment, the techniques he developed enabled others to make important advances in high-pressure science and engineering, such as the ability to synthesize diamonds, which was first done in 1955. Bridgman received the 1946 Nobel Prize in physics for the development of his experimental apparatus, and for the discoveries he made using that apparatus. Bridgman is also known for his writings on the conceptual foundations of physics. Among his works in this area are *The Logic of Modern Physics (1927)* and *The Way Things Are (1959).*

Sir Edward Victor Appleton

*"British physicist Sir Edward Victor Appleton won the Nobel Prize in physics in 1947. Appleton discovered the layer of the ionosphere which reflects high frequency radio waves."*

# Sir Edward Victor Appleton

**Appleton, Sir Edward Victor** (1892-1965), British physicist, who received the Nobel Prize in physics in 1947 for his discovery of the Appleton layer of the ionosphere, which reflects high-frequency radio waves. Appleton was born in Bradford, Yorkshire, England, and educated at Saint John's College, University of Cambridge. He was a captain in the Signal Corps during World War I. From 1918 to 1939 he taught physics at Cambridge and at the University of London, where he began a series of experiments with radio waves that led to his discovery of the Appleton layer. He was secretary of the Department of Scientific and Industrial Research of the British government from 1939 to 1949. This position gave him overall charge of British research during World War II and brought him into close cooperation with American scientists, particularly in connection with the development of radar. For this work, he received the United States Medal of Merit.

Patrick Blackett

*"British physicist Patrick Blackett won the Nobel Prize in physics in 1948. He developed the wilson cloud chamber which he used to study nuclear physics and cosmic radiation."*

# PATRICK BLACKETT

**Blackett, Patrick** (1897-1974), British physicist and Nobel laureate, noted for his contributions to the understanding of cosmic rays. Blackett was born in London and educated at the Royal Naval College and the University of Cambridge. After teaching at various other institutions, he served (1953-65) as professor of physics at the Imperial College of Science and Technology of the University of London. During World War II he was chief adviser on operational research to the British navy. For his discoveries in the fields of nuclear physics and cosmic rays and his work with the cloud chamber, he was awarded the 1948 Nobel Prize in physics. In 1969 he was created a life peer. He wrote *Atomic Weapons* and *East-West Relations (1956)* and *Studies of War (1962).* See *Cosmic Rays; Particle Detectors.*

Hideki Yukawa

*"Japanese physicist Hideki Yakawa won the 1949 Nobel Prize in physics. Based on his research into quantum mechanics and the fields of force affecting elementary particles, he theoretically deduced the existence of mesons, a family of subatomic particles composed of quarks and antiquarks and having intermediate mass."*

# HIDEKI YUKAWA

**Yukawa, Hideki** (1907-1981), Japanese physicist and Nobel laureate, noted for his study of nuclear forces.

Yukawa was born in Tokyo and was educated at the universities of Kyôto and Ôsaka. He became a lecturer in physics at Kyôto University in 1932 and was made professor in 1939. Yukawa also taught (1933-36) at Ôsaka University and was assistant professor there until 1939. He was visiting professor at the Institute for Advanced Studies at Princeton, New Jersey, in 1948 and at Columbia University from 1949 to 1953. Yukawa became (1950) professor emeritus at Ôsaka University and was named (1953) director of the Research Institute for Fundamental Physics at Kyôto University. Yukawa did extensive research in quantum mechanics and the fields of force affecting elementary nuclear particles. In 1935 he theoretically deduced the existence of the meson, for which he was awarded the 1949 Nobel Prize in physics. The existence of the meson was proved in 1936.

Cecil F. Powell

*"British nuclear physicist Cecil F. Powell won the 1950 Nobel Prize in physics. He invented a photographic method for detecting subatomic particles, which led to his discovery of a new particle called the pi-meson, or pion."*

# CECIL FRANK POWELL

**Powell, Cecil Frank** (1903-1969), British nuclear physicist and Nobel Prize winner. A leading proponent of international scientific cooperation, Powell developed photographic methods that greatly enhanced the study of nuclear processes. He won the 1950 Nobel Prize in physics for his photographic methods and for discovering pi-mesons, or pions, a type of cosmic ray (cosmic rays are fast-moving subatomic particles that enter the earth's atmosphere from outer space).

Born in Tonbridge, Kent, England, Powell graduated from Sidney Sussex College of the University of Cambridge in 1925 with a B.S. degree in natural science, and earned his Ph.D. degree in physics in 1927 at the Cavendish Laboratory at Cambridge. In 1928 Powell became research assistant at the University of Bristol, and served as a professor of physics there from 1948 to 1963. In 1963 he was appointed vice chancellor of the University of Bristol, as well as director of the university's H. H. Wills Physics Laboratory, a post from which he retired only months before his death.

Powell became interested in subatomic-particle-detection devices when he studied at the Cavendish Laboratory under Scottish physicist Charles Thomson Rees Wilson, the Nobel Prize-winning inventor of the cloud chamber (a device that tracks ionized particles by producing a trail of water droplets from air saturated with water, and the tool used at the time for detecting subatomic particles). The drawback of the cloud chamber is that it requires a resting period-a luxury seldom afforded a scientist attempting to record the fast-moving tracks of electrically charged particles-each time it is used. As an alternative, Powell began experimenting with photographic emulsions that could be used continuously, but found that none of those available were of adequate quality to record evidence of the particles. He finally persuaded a photographic film company to create a new emulsion that would work. To obtain data on cosmic rays, Powell sent the new emulsion aloft in hydrogen balloons to high altitudes to capture traces of the rays, a project that eventually evolved into a collaborative effort among scientists and nonscientists throughout Europe. These expeditions resulted in Powell's discovery of a new particle, the pi-meson, or pion, which proved to be a cohesive force within the atomic nucleus. These accomplishments earned him the Nobel Prize and marked the beginning of elementary-particle physics.

Increasingly concerned about social problems attached to scientific and technological advances, Powell served from 1961 to 1963 as the chairperson of the Scientific Policy Committee of the European Center for Nuclear Research (CERN, now the European Laboratory for Particle Physics). He was also the founder of the Pugwash Conferences for Science and World Affairs.

Sir John Cockcroft

*"British physicist Sir John Cockcroft won the Nobel Prize in physics in 1951. Cockcroft studied the nature of the atomic nucleus."*

# Sir John Douglas Cockcroft

**Cockcroft, Sir John Douglas** (1897-1967), English nuclear physicist, who shared the 1951 Nobel Prize for physics for designing the first particle accelerator (an instrument used to increase the energy of subatomic particles) and inducing the first artificial nuclear transformation of an element.

Born in Todmorden, Yorkshire, Cockcroft was educated at the Manchester College of Technology and the University of Cambridge, where he studied mathematics and electrical engineering. After graduating from Cambridge in 1924, he worked at the Cavendish Laboratory under English physicist Ernest Rutherford and collaborated with Russian physicist Peter Kapitza to design and develop powerful electromagnets. Cockcroft's interests soon turned to nuclear physics and the challenge of accelerating atomic particles in an electric field. During the 1920s the only particles used for bombarding and breaking down the atomic nucleus (a process known as atom smashing) were alpha particles emitted by naturally radioactive elements. Rutherford had thoroughly exploited this approach, and Cockcroft believed it necessary to find particles of still higher energies to improve the process.

In 1929 Cockcroft and colleague Ernest T. S. Walton developed a type of particle accelerator they called a *voltage multiplier*. It was a machine that could build up voltages and accelerate protons to energies higher than those of natural alpha particles. In 1932 Cockcroft and Walton bombarded lithium with these protons and produced alpha particles. In this reaction they managed to combine lithium and hydrogen to form helium; it was the first nuclear reaction induced by artificially accelerated particles.

In 1939 Cockcroft became a professor at the University of Cambridge. During World War II (1939-1945) he was appointed director of Britain's air defense research and contributed to the development of that nation's radardefense system. He was also involved in the development of nuclear energy and was director of the Canadian National Research Council. From 1946 to 1958 he served as the first director of the British Atomic Energy Research Establishment in Harwell, England. He was knighted for his accomplishments in 1948, and in 1951 he and Walton shared the Nobel Prize for physics. Cockcroft was appointed master of Churchill College in Cambridge in 1960 and received the Atoms for Peace award in 1961.

Ernest T. S. Walton

*"Irish physicist Ernest T. S. Walton won the 1951 Nobel Prize in physics. He invented the particle accelerator and was the first to split atoms and to successfully transmutate elements."*

# Ernest Thomas Sinton Walton

**Walton, Ernest Thomas Sinton** (1903-1995), Irish physicist and Nobel Prize winner. Walton advanced the field of nuclear physics significantly, as the first scientist to create a particle accelerator (a device to smash atoms or subatomic particles together at high speeds), the first to split atoms, and the first to transform one element into another. He and British physicist Sir John Douglas Cockcroft won the 1951 Nobel Prize in physics for their atomic research with particle accelerators.

Born in Dungarvan, Ireland, Walton received a B.S. degree in physics and mathematics from Trinity College (the University of Dublin) in 1926, and an M.S. degree in physics from Trinity in 1927. He earned his Ph.D. degree from the University of Cambridge in England in 1930. Walton began his professional career in 1927 at Cambridge's Cavendish Laboratory, as research assistant to British physicist Sir Ernest Rutherford. Walton held this position until 1934, when he accepted a professorship at Trinity College, where he remained until his retirement in 1974.

While working for Rutherford at Cambridge, Walton embarked on research that would bring him wide recognition. He built a linear particle accelerator that was to prove a prototype for subsequent particle accelerators. In 1932, in collaboration with Cockcroft, Walton used this device to bombard lithium atoms, generating enough force to transform each lithium nucleus into two helium nuclei. This marked the first time that human means proved successful in the transmutation of elements (changing one element into another). Walton later succeeded in accelerating protons to such a high velocity that they were able to penetrate atomic nuclei of light elements and start nuclear reactions. He was especially interested in the relations created during this process between the energies of the protons before they hit the nuclei and those of the created nuclear particles. These experiments demonstrated that atomic nuclei contain enormous energies. The research also provided the first experimental confirmation of equations for the equivalence of mass and energy by American physicist Albert Einstein.

Walton's discoveries have had a great influence on the field of nuclear physics in the latter half of the 20th century. Physicists in laboratories in many countries continue to use his methods for accelerating charged particles to produce nuclear reactions.

Felix Bloch

*"Swiss-born American physicist Felix Bloch won the Nobel Prize in physics 1952. Bloch won the prize for his development of high-precision methods in nuclear magnetism and for discoveries stemming from these methods."*

## FELIX BLOCH

**Bloch, Felix** (1905-1983), Swiss-born American physicist, educator, and cowinner of the 1952 Nobel Prize for physics. Bloch shared the Nobel Prize with American physicist Edward Mills Purcell for their development of a new method for the precise measurements of the strength of the magnetic field of the atomic nucleus, called nuclear magnetic resonance (NMR). A number of important applications have come from NMR, including magnetic resonance imaging (MRI). MRI produces detailed internal images of the human body, which helps physicians diagnose disease and injuries.

Edward M. Purcell

*"American physicist Edward M. Purcell won the 1952 Nobel Prize in physics. Purcell developed new methods for measuring nuclear magnetism, enabling researchers to determine atomic and molecular structures more rapidly than was previously possible."*

# EDWARD MILLS PURCELL

**Purcell, Edward Mills** (1912-1997), American physicist, educator, and cowinner of the 1952 Nobel Prize for physics. Purcell and American physicist Felix Bloch shared the Nobel Prize for their contributions in developing new methods to measure the strength of the magnetic field in an atom's nucleus, called nuclear magnetic resonance (NMR). A number of important applications came from NMR, including magnetic resonance imaging (MRI). MRI produces detailed images of the human body, which helps physicians diagnose injuries and diseases.

Born in Taylorville, Illinois, Purcell graduated with a B.S. degree in electrical engineering from Purdue University in 1933. After further studies at a technical university in Karlsruhe, Germany, he entered Harvard University and earned his Ph.D. degree in physics in 1938. Upon graduation, Purcell served as an instructor at Harvard until 1940. During World War II (1939-1945) he worked in an advanced radiation laboratory at the Massachusetts Institute of Technology (MIT), where he met Bloch. Purcell rejoined the faculty at Harvard in 1945 as an associate professor and became a full professor in 1949. He remained there until his retirement in 1980.

In 1945 Purcell and colleagues at Harvard successfully used NMR to measure how much electromagnetic radiation of a specific frequency is absorbed by an atomic nucleus that is placed in a strong magnetic field. This method helps to reveal atomic and molecular structures. At the same time, a research group, headed by Bloch, then a professor at Stanford University, made similar observations.

Scientists, researchers, and the general public continue to benefit from Purcell's discoveries. NMR revolutionized the field of chemistry and has become the most important spectroscopic technique in chemistry and biology. Chemists use NMR instruments to check the moisture content of food and check the quality of drugs and medicines. NMR also helps researchers probe the nature of ribonucleic acid (RNA) and deoxyribonucleic acid (DNA), the nucleic acids that are the building blocks of human life.

Purcell also made major contributions to the study of radio astronomy, astrophysics, and biophysics. In 1951, using NMR, he discovered that hydrogen atoms in interstellar space emitted microwaves with a wavelength of 21 cm (8.27 in). This discovery enabled astronomers to make observations in a new way that does not require light.

Frits Zernike

*"Dutch physicist Frits Zernike won the 1953 Nobel Prize in physics. He greatly advanced the study to microbiology with his invention of the phase microscope, which reveals details that are indiscernible with a conventional microscope."*

# FRITS ZERNIKE

**Zernike, Frits** (1888-1966), Dutch physicist and Nobel Laureate. Zernike invented the phase-contrast, or phase, microscope, a microscope that can show tiny differences in the way a transparent specimen bends light .The phase-contrast microscope is especially useful in studying living tissue. For that invention, Zernike was awarded the 1953 Nobel Prize in physics.

The son of two mathematics teachers, Zernike was born in Amsterdam, the Netherlands. He earned his Ph.D. degree in physics in 1915, at the University of Amsterdam. In 1913 he accepted a position as assistant to the Dutch astronomer Jacobus Kapteyn at the University of Groningen, where Zernike became a lecturer in theoretical physics two years later. In 1920 he was promoted to full professor of theoretical physics, and in 1941 expanded his chair to include mathematical and technical physics and theoretical mechanics. Later in his career, Zernike was briefly affiliated with Johns Hopkins University in Baltimore, Maryland.

Conventional optical microscopes are inadequate for viewing the detail in living specimens, particularly if the specimens are transparent-the detail cannot be seen unless the tissue is stained, which often kills the tissue. The problem is that in a microscope image, there are variations in the phase of light (determined by the path that light travels) that the eye cannot detect.

Zernike discovered that the effects due to changes in the optical path can be transformed into changes in light intensity, which the eye can detect. To accomplish this transformation, he invented a microscope that utilized a diaphragm that funneled light into a cone focused on the specimen, and a *diffraction plate,* or *phase plate*, inserted between the two components of the objective lens (the microscope lens that forms an image of an object). Direct light passes through the groove in the plate, and diffracted light passes outside the groove in the plate. Manipulation of the direct and diffracted light creates an optical-path difference that results in greater light intensity (the two beams interfere constructively), revealing detail indiscernible with a conventional microscope. Although the phase-contrast microscope greatly advanced microbiology by making possible the detailed examination of biological and medical specimens, Zernike spent many years trying to find a company that would manufacture his invention.

Max Born

*"German-British physicist Max Born won the Nobel Prize in physics in 1954. Born contributed to the understanding of quantum theory."*

## MAX BORN

**Born, Max** (1882-1970), German-British physicist and Nobel laureate, born in Breslau (now Wroc³aw, Poland), and educated at the universities of Breslau, He--idelberg, Zürich, Göttingen, and Cambridge. In 1921, after teaching successively at the universities of Göttingen, Berlin, and Frankfurt, he was appointed professor of theoretical physics at Göttingen. He escaped to Britain in 1933, a refugee from Germany, and acquired British citizenship in 1939. During his first three years in England, he conducted research at the University of Cambridge. He was Tait Professor of Natural Philosophy at the University of Edinburgh from 1936 to 1953. An outstanding theoretical physicist noted for his fundamental contributions in quantum theory, Born shared the 1954 Nobel Prize in physics with the German physicist Walter Bothe. His works include Einstein's Theory of Relativity (1922), Atomic Physics (1935), Natural Philosophy of Cause and Chance (1949), Physics and Politics (1962), and My Life and My Views (1968).

Walter Bothe

*"German physicist, mathematician and chemist Walther Bothe won the Nobel Prize in physics in 1954. His work involved particle detection and the study of interactions between elementary particles."*

# WALTHER WILHELM GEORG BOTHE

**Bothe, Walther Wilhelm Georg** (1891-1957), German physicist, mathematician, chemist, and Nobel Prize winner. Bothe developed the coincidence method of particle detection and used it to study interactions between elementary particles. He shared the 1954 Nobel Prize in physics with German-born British physicist Max Born.

Born in Oranienburg, Germany, Bothe attended the University of Berlin from 1908 to 1912. He received his doctorate in 1914 under the German physicist Max Planck for the study of the molecular theory of refraction, reflection, scattering, and absorption of light rays. Bothe served in the Germany Army during World War I (1914-1918) and was captured and held as a prisoner of war in Russia from 1915 to 1920. While he was a prisoner, he managed to keep studying theoretical physics and learn Russian. In 1920 he accepted an invitation from Hans Geiger, a fellow German physicist, to work at the radioactivity laboratory of the Physikalisch-Technische Reichsanstalt (State Physical-Technical Institute). Bothe went on to teach physics at the University of Berlin from 1920 to 1931, and at the University of Giessen from 1931 to 1934. In 1934 he became director of the Max Planck Institute at Heidelberg, where he remained until his death.

Bothe's international recognition began with his development of mathematical expressions for the scattering patterns of various rays and particles. This research served as foundation for his studies on collisions between X rays and electrons in matter, in which the electrons are knocked away from their atoms and the X rays are scattered. Earlier physicists had theorized that individual collisions did not necessarily conserve momentum and energy-that is, the momentum and energy of the incoming X ray did not have to equal the combined momentum and energy of the electron and the scattered X ray after the collision. This idea was contrary to the fundamentals of classical physics.

Geiger and Bothe believed they could test this theory experimentally by using electrical counters to examine particle emissions. The most widely used electrical counter was the Geiger-Müller tube.

With his knowledge of the tube, Bothe invented a system in 1925 in which two counters were connected to a common amplifier. The device registered pulses only when particles triggered both counters simultaneously. This new process, called the coincidence method, allowed Bothe and Geiger to study the coincidences between the scattered X ray and the recoiling electron, leading them to discover small-scale conservation of energy and momentum, thus refuting the quantum theory of radiation.

Bothe continued his work in experimental and theoretical physics, working on cosmic rays (high-energy, fast moving subatomic particles that come through the earth's atmosphere from outer space), discovering a new type of radiation later found to be the neutron, and working on the German government's radiation and nuclear energy projects.

Polykarp Kusch

*"American physicist Polykarp Kusch won the 1955 Nobel Prize in physics. He helped establish the magnetic field properties of the electron."*

# POLYKARP KUSCH

**Kusch, Polykarp** (1911-1993), American physicist and Nobel laureate, born in Blan–kenburg, Germany. He went to the United States in 1912, and was educated at Case Institute of Technology and the University of Illinois. Kusch worked mainly at Columbia University and the University of Texas at Dallas. He shared the 1955 Nobel Prize for physics with the American physicist Willis Eugene Lamb Jr. for establishing the magnetic moment of the electron.

Willis E. Lamb, Jr.

*"American physicist Willis E. Lamb, Jr. won the 1955 Nobel Prize in physics. He showed that the two energy states of the hydrogen atom were slightly different, a phenomenon now called the Lamb shift."*

# Willis Eugene Lamb, Jr.

**Lamb, Willis Eugene, Jr.** (1913- ), American physicist and Nobel Prize winner. Lamb made groundbreaking discoveries about the energy states of the hydrogen atom. Lamb and German-born American physicist Polykarp Kusch shared the 1955 Nobel Prize in physics for their independent research.

Born in Los Angeles, California, Lamb completed his B.S. degree in chemistry at the University of California, Berkeley, in 1934 and stayed on to receive his Ph.D. degree in physics in 1938. He went on to be an instructor (from 1938 to 1948) and professor (from 1948 to 1951) in physics at Columbia University in New York City. He was a professor of physics at Stanford University in California from 1951 to 1956, at the University of Oxford in England from 1956 to 1962, and at Yale University in New Haven, Connecticut, from 1962 to 1974. In 1974 Lamb became a professor of physics at the University of Arizona in Tucson, where he remains today.

Lamb is best known for his accurate studies of the hydrogen spectrum's hyperfine structure, which reveals that there can be tiny differences in energy between two atoms in the same energy level. In 1947, through sophisticated measurements, he showed that the two possible energy states of hydrogen differed by a very small amount. This contradicted earlier predictions by English physicist Paul Dirac that these energy states were equal. Lamb's discovery is called the Lamb shift. The shift required physicists to revise their understanding of the interaction of the electron with electromagnetic radiation.

Lamb has also contributed significantly to the study of theoretical physics, atomic and nuclear structure, microwave spectroscopy, and maser and laser physics. In his current position as the Regents Professor of Physics and Optical Sciences at the University of Arizona, Lamb continues to conduct research on theoretical atomic physics and quantum optics.

William Bradford Shockley

*"American physicist William Shockley won the 1956 Nobel Prize in physics. He was awarded the prize for his research on semiconductors, which led to the development of the transistor."*

# William Bradford Shockley

**Shockley, William Bardford** (1910-1989), American physicist, Nobel laureate, and coinventor of the transistor, Shockley was born in London and worked at Bell Telephone Laboratories from 1936 to 1956, when he become director of the Shockley Transistor Crop in Palo Alto, California. He lectured at stanford university beginning in 1958 and become professor of engineering science in 1963. His research on semiconductors led to the development of the transistor in 1948. For this research he shared the Noble prize in physics with his associates John Bardeen and Walter H. Brattain. More recently he published several controversial papers arguing that intelligence is primarily hereditary.

Walter Houser Brattain

*"American physicist Walter Houser Brattain won the Nobel Prize in physics in 1956. Brattain worked on the team that developed the transistor and the semiconductor."*

# WALTER HOUSER BRATTAIN

**Brattain, Walter Houser** (1902-1987), American physicist and Nobel laureate, born in Xiamen (Amoy), China. After working as a physicist in the radio division of the National Institute of Standards and Technology, in 1929 he joined the staff of Bell Telephone Laboratories. While working at Bell, Brattain and the American physicists William Shockley and John Bardeen developed a small electronic device called the transistor. First announced in 1948, the transistor was perfected by 1952 for commercial use in portable radios, hearing aids, and other devices. For his work on semiconductors and discovery of the transistor effect, Brattain shared the 1956 Nobel Prize in physics with Shockley and Bardeen.

John Bardeen

*"American physicist John Bardeen won the Nobel Prize twice in physics one in 1956 and other in 1972. Bardeen helped develop the transistor, an electronic device that performed many of the functions of the vacuum tube."*

## JOHN BARDEEN

**Bardeen, John** (1908-1991), American physicist and Nobel laureate, born in Madison, Wisconsin, and educated at the University of Wisconsin and Princeton University. As a research physicist (1945-1951) at the Bell Telephone Laboratories in Murray Hill, New Jersey, he was a member of the team that developed the transistor, a tiny electronic device capable of performing most of the functions of the vacuum tube. For this work, he shared the 1956 Nobel Prize in physics with two colleagues, the American physicists William Shockley and Walter H. Brattain. Meanwhile he had joined (1951) the faculty of the University of Illinois. In 1972 he shared the Nobel Prize in physics with the American physicists Leon N. Cooper and John R. Schrieffer for the development of a theory to explain superconductivity, the disappearance of electrical resistance in certain metals and alloys at temperatures near absolute zero. Bardeen thus became the first scientist to win two Nobel Prizes in the same category.

Tsung-Dao Lee

*"American Nuclear physicist Tsung-Dao Lee won the 1957 Nobel Prize in physics. He disproved the law of conservation of parity by studying weak nuclear reactions."*

## TSUNG-DAO LEE

**Lee, Tsung-Dao** (1926-), Chinese-born American nuclear physicist and Nobel laureate, noted for his work in quantum mechanics. He was born in Shanghai and educated at the Chekiang and National Southwest Associated universities in China and at the University of Chicago. He has held positions at the University of California, the Institute for Advanced Study at Princeton, New Jersey, and Columbia University. With his associate Chen Ning Yang, he proved experimentally that the law of conservation of parity did not hold true for weak nuclear reactions. For this discovery the two men were awarded the 1957 Nobel Prize in physics.

Chen Ning Yang

*"Chinese-born American theoretical physicist Chen Ning Yang won the 1957 Nobel Prize in physics. Much of his work centered on the nature and behaviour of elementary particles and, through his experiments, he disproved one of the basic laws of quantum mechanics."*

# CHEN NING YANG

**Yang, Chen Ning** (1922-), American theoretical physicist and Nobel laureate, best known for his studies of the nature and behavior of elementary particles. Yang was born in Hefei (Ho-fei), China, and educated at the Southwest Associated University in China and at the University of Chicago. He taught physics at the University of Chicago from 1948 to 1949, when he was invited to do his research at the Institute for Advanced Study, in Princeton, New Jersey. He was made a permanent member of the institute in 1952 and full professor in 1955. In 1965 he was named Albert Einstein Professor of Physics at the State University of New York at Stony Brook.

Yang is noted for his work in the field of quantum theory. With his associate Tsung-Dao Lee, he proved experimentally that one of the basic quantum-mechanics laws, called the conservation of parity, is violated in the so-called weak nuclear reactions, those nuclear processes that result in the emission of beta or alpha particles. In recognition of this achievement Yang and Lee shared the 1957 Nobel Prize in physics. Yang wrote *Elementary Particles (1962).*

Pavel Cherenkov

*"Russian physicist Pavel Cherenkov won the Nobel Prize in physics in 1958. Cherenkov discovered that certain liquids will glow when irradiated with gamma rays."*

# Pavel Alekseyevich Cherenkov

**Cherenkov, Pavel Alekseyevich** (1904-1990), Russian physicist and Nobel laureate. In 1932 he began to study the luminescence given off by certain liquids when irradiated by gamma rays and, in 1934, discovered a phenomenon now known as the Cherenkov effect. This phenomenon (also called Cherenkov radiation) is the emission of a bluish light from a liquid when electrons or other charged atomic particles move through the liquid with a velocity greater than that of light in the same medium. That is, the velocity of light, c, is reduced to the velocity c/n in a dialectric medium, where n is the refractive index of the medium. If a particle enters the medium at a velocity very nearly that of c, its velocity can then exceed c in that medium. Radiation detectors called Cherenkov counters make use of this phenomenon. In 1958 Cherenkov shared the Nobel Prize in physics with Ilya M. Frank and Igor Y. Tamm for the discovery of the Cherenkov effect.

Il'ja M. Frank

*"Russian physicist I. M. Frank won the Nobel Prize in physics in 1958. Frank advanced nuclear physics and the study of cosmic rays."*

## Il'ja Mikhailovich Frank

**Frank, I. Mikhailovich** (1908-1990), Russian physicist and cowinner of the 1958 Nobel Prize in physics for his interpretation of the Cherenkov effect. This phenomenon occurs when high-energy, charged particles travel through a medium, such as water or plastic, at a speed greater than the speed of light in the same medium. The result is the emission of bluish light. His research greatly advanced nuclear physics and the study of cosmic rays (protons and atomic nuclei from outer space). Frank shared the 1958 Nobel Prize with Soviet physicists Pavel Cherenkov and Igor Tamm.

Igor Tamm

*"Soviet physicist Igor Tamm won the 1958 Nobel Prize in physics. One of the great theoretical physicist, much of his ground breaking research concerned the behaviour of light, including his theoretical explanation of the Cherenkov effect."*

# IGOR YEVGENYEVICH TAMM

**Tamm, Igor Yevgenyevich** (1895-1971), Soviet physicist and Nobel laureate, who based his work on the Einstein theory of relativity and on quantum mechanics. Tamm was born in Vladivostok and educated at Moscow University. Considered one of the outstanding theoretical physicists in the world, he developed (1924-30) the quantum theory of acoustical vibrations and the scattering of light in solid bodies, as well as the theory of interactions of light with electrons. In 1933 he theorized on the existence of surface states (Tamm's levels) of electrons in semiconductors. In 1937 he and Ilya Frank worked out a theoretical interpretation for the Cherenkov effect. Tamm suggested (1950) the use of electric charges in ionized gases as a means of obtaining controlled thermonuclear power. For their work on the Cherenkov effect, Tamm, Frank, and Pavel Cherenkov shared the 1958 Nobel Prize in physics.

Owen Chamberlain

*"American physicist Owen Chamberlain won the Nobel Prize in physics in 1959. Chamberlain confirmed the existence of the anti-proton, a form of antimatter."*

# OWEN CHAMBERLAIN

**Chamberlain, Owen** (1920- ), American physicist and Nobel laureate. For their collaborative discovery of the antiproton Chamberlain and Italian-born American physicist Emilio Gino Segrè shared the 1959 Nobel Prize in physics.

Born in San Francisco, California, Chamberlain completed his undergraduate studies at Dartmouth College, receiving a B.S. degree in physics in 1941. From 1942 to 1946 he served as a researcher on the Manhattan Project, which developed the atomic bomb during World War II (1939-1945). From 1947 to 1949 he worked at the Argonne National Laboratory near Chicago, Illinois, obtaining a Ph.D. degree in physics at the University of Chicago in 1949. He then joined the faculty of the University of California at Berkeley, becoming a professor of physics in 1958, a position he held until his retirement in 1989.

Chamberlain's work in nuclear physics began with close investigation of subatomic particles, or particles that make up atoms. His part in the atomic-bomb project led him to study alpha-particle decay, neutron diffraction, and high-energy nuclear reactions. In 1955 Chamberlain, with Emilio Segrè, discovered the antiproton, a form of antimatter. Later Chamberlain also confirmed the existence of the antineutron.

Emilio Segre

*"Italian- American nuclear physicist Emilio Segrè won the 1959 Nobel Prize in physics. He was awarded the prize for discovering the antiproton."*

# EMILIO GINO SEGRÈ

**Segrè, Emilio Gino** (1905-1989), Italian American nuclear physicist and Nobel laureate, who was born in Rome and educated at the University of Rome. With American physicist Owen Chamberlain, he detected the antiproton in 1955 using the betatron particle accelerator at Lawrence Berkeley National Laboratory. For this they shared the 1959 Nobel Prize in physics. Segrè also took part in the discovery of the elements astatine and technetium and the isotope plutonium-239.

Donald Glaser

*"American physicist Donald Glaser won the 1960 Nobel Prize in physics. He invented the bubble chamber, a particle detector that indicates the paths of high-energy particles."*

# DONALD ARTHUR GLASER

**Glaser, Donald Arthur** (1926- ), American physicist and winner of the 1960 Nobel Prize in physics for his invention of the bubble chamber, a device for detecting high-energy particles. With this device, Glaser contributed greatly to the understanding of atomic function and provided the technology for the discovery of new atomic particles.

Glaser was born in Cleveland, Ohio, the son of Russian immigrants. An accomplished violinist, Glaser became a member of a symphony orchestra at age sixteen and studied composition at the Cleveland Institute of Music. He received his B.S. degree in mathematics and physics from Cleveland's Case Institute of Technology (now Case Western Reserve University) in 1946, and his Ph.D. degree from the California Institute of Technology in 1950. In 1949 Glaser accepted a position as instructor at the University of Michigan, where he began his bubble chamber experiments. In 1959 Glaser became a professor of physics at the University of California, Berkeley.

In creating the bubble chamber, Glaser built on the work of Nobel Prize winners C. T. R. Wilson, the inventor of the cloud chamber (a device that exposes charged particles by producing a trail of water droplets from air saturated with water), and C. F. Powell, who developed photographic techniques to capture images of charged nuclear particles on film. Both the cloud chamber and the emulsion technique were limited to the detection of low-energy nuclear particles. Glaser wanted to build a device that could reveal high-energy particles. He experimented with soda, seltzer, and beer without success. But he found that nuclear particles left a trail of bubbles as they moved through superheated ether (a colorless, flammable liquid). Using the bubble chamber, scientists have discovered new high-energy atomic particles. They have also used it to track neutral atomic particles advancing their understanding of the mass, lifetime, and decay of these particles.

Robert Hofstadter

*"American physicist Robert Hofstadter won the 1961 Nobel Prize in physics. He won the award for his research into the size and shape of protons and neutrons."*

## ROBERT HOFSTADTER

**Hofstadter, Robert** (1915-1990), American physicist and Nobel laureate, born in New York City. Hofstadter received his Ph.D. degree from Princeton University in 1938 and joined the faculty of Stanford University in 1950. Using the linear accelerator at the Stanford Linear Accelerator Center and a scattering machine of his own design, Hofstadter was able to measure with great precision the size and shape of the proton and neutron. Both were found to consist of a positively charged, dense, pointlike core surrounded by two intermingling layers of meson clouds Elementary Particles.. Hofstadter shared the 1961 Nobel Prize in physics with Rudolf Ludwig Mössbauer.

Rudolf Mossbauer

*"German physicist Rudolf Mössbauer won the 1961 Nobel Prize in physics. He discovered a property of certain radioactive atoms, called the Mössbauer effect, that enables precise measurements of gravitational magnetic, and electrical effects."*

# RUDOLF LUDWIG MÖSSBAUER

**Mössbauer, Rudolf Ludwig** (1929-), German physicist and Nobel laureate. He is known for his discovery of the Mössbauer effect. Mössbauer was born in Munich and received his Ph.D. degree in physics from the Technical Institute in Munich in 1958. In 1960 he came to the United States, joining the staff of the California Institute of Technology.

In 1953 Mössbauer began to investigate the absorption of gamma rays in matter, and in 1955 he first observed the phenomenon now known as the Mössbauer effect, for which he won the 1961 Nobel Prize in physics. When a gamma ray is emitted from a radioactive atom, a recoil effect usually occurs. Mössbauer discovered that in certain radioactive substances in which the atoms are held in a tight crystalline structure, no recoil occurs and the gamma rays are emitted at a frequency corresponding to the exact difference between the nuclear ground-state energy and the excited-state energy. If such a gamma ray strikes an atom of the same element held in a similarly tight crystalline structure, another gamma ray of exactly the same frequency may be emitted. The Mössbauer effect enables extremely precise measurements to be made of gravitational, magnetic, and electrical effects.

Lev D. Landau

*"Soviet theoretical physicist Lev D. Landau won the 1962 Nobel Prize in physics. He developed theories to describe the behaviour of superfluid liquid helium at extremely low temperatures."*

## LEV DAVIDOVICH LANDAU

**Landau, Lev Davidovich** (1908-1968), Soviet theoretical physicist and Nobel lau--reate, noted chiefly for his pioneer work in low-temperature physics (cryogenics). He was born in Baku, and educated at the universities of Baku and Leningrad. In 1937 Landau became professor of theoretical physics at the S. I. Vavilov Institute of Physical Problems in Moscow. His development of the mathematical theories that explain how superfluid helium behaves at temperatures near absolute zero earned him the 1962 Nobel Prize in physics. His writings on a wide variety of subjects relating to physical phenomena include some 100 papers and many books, among which is the widely known nine-volume Course of Theoretical Physics, published in 1943 with Y. M. Lifshitz. In January 1962, he was gravely injured in an automobile accident; he was several times considered near death and suffered a severe impairment of memory. By the time of his death he had been able to make only a partial recovery.

Eugene Paul Wigner

*"Hungarian-born American physicist Eugene Paul Wigner won the 1963 Nobel Prize in physics. In addition to developing the nuclear reactor, he made important contributions, to the understanding of the structure of the atomic nucleus and elementary particles."*

## EUGENE PAUL WIGNER

**Wigner, Eugene Paul** (1902-1995), American physicist and Nobel for his work on quantum physics and the development of nuclear reactors. Wigner was born in Budapest, Hungary. He joined the faculty of Pricneton University in 1930 and become a U.S. citizen in 1937. He was one of five scientists who informed President Franklin D. Roosevelt in 1939 of the possible military use of atomic energy, and during world war II he design plutonium reactors. He shared the 1963 Nobel prize in physics for his work in elucidating the structure of the atomic nucleus and his development of quantum mechanics theory concerning the nature of the proton and neutron.

Maria Goeppert-Mayer

*"German-American physicist Maria Goeppert-Mayer won the Nobel Prize in physics in 1963. She studied the structure of the atomic nucleus."*

# MARIA GOEPPERT-MAYER

**Goeppert-Mayer, Maria** (1906-1972), German American physicist and Nobel laureate, best known for her study of nuclear structure. Goeppert-Mayer was born in Poland and educated at the University of Göttingen. In 1931 she married the American physicist Joseph E. Mayer and went with him to the United States, becoming a U.S. citizen in 1933. She taught at several institutions before joining (1960) the faculty of the University of California at San Diego. She shared the 1963 Nobel Prize in physics and was cited by the Nobel committee for her independent work in the late 1940s, in which she demonstrated that the atomic nucleus has a structure containing successive proton-neutron shells held together by complex forces.

J. Hans Jensen

*"German physicist J. Hans Jensen won the 1963 Nobel Prize in physics. He was awarded the prize for his work in developing a shell model of the atomic nucleus."*

# J. Hans Daniel Jensen

**Jensen, J. Hans Daniel** (1907-1973), German physicist and Nobel Prize winner. Jensen won the 1963 Nobel Prize in physics for advancing understanding of atomic structure (see Atom) by developing the concept that the particles making up the nucleus of atoms arrange themselves in shells. The prize was shared by American physicist Maria Goeppert-Mayer, who independently developed a shell model of the atomic nucleus, and by Hungarian-born American physicist Eugene Paul Wigner, whose work provided the background needed by Goeppert-Mayer and Jensen.

Born in Hamburg, Jensen attended the Universities of Freiburg and Hamburg, earning his Ph.D. degree in physics from the University of Hamburg in 1932. Upon his graduation, he became a professor at the University of Hamburg, holding the position until 1941. He spent the next seven years as a professor at the Hanover Institute of Technology. In 1949 Jensen undertook a professorship at the University of Heidelberg, a position he held until his retirement in 1969. For periods from 1951 to 1953 and again in 1961 he left Heidelberg to lecture in the United States.

Jensen made a striking advance in the correlation of nuclear properties when he helped develop the shell model of the atomic nucleus. The theory, put forth by Jensen in 1949, was that protons and neutrons are arranged in shells or spherical layers in atoms, as are the atom's electrons. This understanding helped Jensen and Goeppert-Mayer make subsequent discoveries concerning shell structure. The two physicists had postulated this structure independently, but decided to work together to further explore it. With Goeppert-Mayer, Jensen found evidence that atomic nuclei with 2, 8, 20, 28, 50, 82, or 126 nucleons (protons or neutrons) show unusual stability. The integers in this context are known as "magic numbers." Jensen's assistance led Goeppert-Mayer to later demonstrate the theoretical basis for a shell model of atomic nuclei, with each magic number corresponding to a completed nuclear shell. In 1955 Jensen and Goeppert-Mayer coauthored the book *Elementary Theory of Nuclear Shell Structure,* which chronicles their discoveries. Jensen also studied other aspects of atomic physics and the interactions between subatomic particles.

Nicolay Basov

*"Russian physicist Nikolay Basov won the Nobel Prize in physics in 1964. Basov's research led to the development of the maser and the laser."*

# NICOLAY GENNADIYEVICH BASOV

**Basov, Nicolay Gennadiyevich** (1922-2001), Soviet physicist and Nobel Laureate. Basov helped to develop both the laser and the maser, for which he shared the 1964 Nobel Prize in physics with Soviet physicist Aleksandr Mikhailovich Prokhorov and American physicist Charles Hard Townes.

Born in Novaya Usman', Russia, Basov graduated from the Moscow Engineering and Physics Institute in 1950 with a degree equivalent to an M.S. degree in physics. He obtained a Ph.D. degree from the Lebedev Institute of Physics at the Soviet Academy of Sciences in Moscow in 1956. In 1950 he became a researcher at the Lebedev Institute, where he has held various directorial duties since 1958.

Basov, together with his teacher Prokhorov, conducted groundbreaking research in quantum mechanics, which concerns the behavior of atoms at different energy levels. They first deduced that quantum mechanics permits the amplification of microwaves and light waves by inducing atoms to release energy. This helped them construct the theoretical basis of the process now called microwave amplification by stimulated emission of radiation, or, more commonly, maser. The maser quickly found many applications for its ability to send strong microwaves in any direction and resulted in improvements in radar. The maser also provided the basis for an atomic clock that was far more accurate than any mechanical timepiece ever invented. Basov later helped develop the visible-light maser, or laser (Light Amplification by Stimulated Emission of Radiation), which delivers infrared or visible light instead of microwaves. Both the maser and the laser can collect and amplify energy waves hundreds of times. They can also produce a beam with almost perfectly parallel light waves and little or no interference or static.

Aleksandr Prokhorov

*"Russian physicist Aleksandr Prokhorov won the 1964 Nobel Prize in physics. He was awarded the prize for his basic research into experimental physics, which led to the invention of the maser and the laser."*

# ALEKSANDR MIKHAILOVICH PROKHOROV

**Prokhorov, Aleksandr Mikhailovich** (1916-2002), Australian-born Soviet physicist and Nobel laureate. Prokhorov helped to develop both the laser and the maser, for which he shared the 1964 Nobel Prize in physics with Soviet physicist Nikolay Gennadiyevich Basov and American physicist Charles Hard Townes.

Prokhorov was born in Atherton, Australia, where his family had fled from Russia in 1911. The family returned to the Union of Soviet Socialist Republics (USSR) in 1923, and Prokhorov graduated from Leningrad University with a B.S. degree in physics in 1939. Shortly thereafter, World War II (1939-1945) interrupted Prokhorov's career. He served in the Soviet Army from 1941 to 1944 and in 1948 obtained his Ph.D. degree in physics at the Soviet Academy of Sciences in Moscow. Two years later, he joined the research staff at the Lebedev Institute of Physics at the Soviet Academy of Sciences in Moscow. He soon became an administrator and is still a member of the staff today.

Prokhorov, together with Basov, conducted groundbreaking research in quantum mechanics which concerns the behavior of atoms at different energy levels. They first deduced that manipulating quantum energies might permit them to amplify microwaves and light waves. They then constructed the theoretical basis of a process now called microwave amplification by stimulated emission of radiation, or maser. The maser quickly found many applications for its ability to send strong microwaves in any direction and resulted in improvements in radar. The maser also provided the basis for an atomic clock that was far more accurate than any mechanical timepiece ever invented. Prokhorov later helped develop the visible-light maser, or laser (light amplification by stimulated emission of radiation), which delivers infrared or visible light instead of microwaves. Both the maser and laser can collect and amplify energy waves hundreds of times. They can also produce a beam with almost perfectly parallel light waves and little or no interference or static. Later in his career, Prokhorov further investigated the interactions of laser radiation with matter.

Charles Townes

*"American physicist Charles Townes won the 1964 Nobel Prize in physics. He made fundamental contributions in quantum theory and significantly improved radar technology."*

# CHARLES HARD TOWNES

**Townes, Charles Hard** (1915- ), American physicist and Nobel laureate. Townes made important contributions to the field of quantum theory and significantly improved radar technology. For his fundamental work in the field of quantum electronics, Townes was awarded the 1964 Nobel Prize in physics, which he shared with Soviet physicists Nikolay Basov and Aleksandr Prokhorov.

Born in Greenville, South Carolina, Townes received his B.S. degree in physics and modern languages from Furman University in 1935. He went on to obtain his M.S. degree in physics at Duke University in 1936 and his Ph.D. degree in physics at the California Institute of Technology (Caltech) in 1939. From 1939 to 1947 he worked at Bell Telephone Laboratories. From 1948 to 1961 he taught physics at Columbia University, and from 1961 to 1967 he taught at the Massachusetts Institute of Technology (MIT). In 1967 he became a professor at the University of California at Berkeley, a position he still holds.

At Columbia, Townes studied radar technology, which involves emitting and receiving microwaves-the wavelength of electromagnetic radiation that falls between infrared waves and radio waves. During his studies, he recognized the need for a device that would generate microwaves in great intensity, and in 1951 he came upon the idea of producing this energy by manipulating molecules rather than electronic circuits. Townes immediately thought of the ammonia molecule, which had a vibrating frequency and two levels of energy that might lend themselves to producing microwaves. He hypothesized that he could get ammonia molecules "excited" by pumping energy into them through heat or electricity, after which he would expose them to a weak beam of microwaves. Townes knew that molecules so treated would emit their own energy in microwaves, which would strike other molecules and cause them to give up their energy. He hoped that the very feeble incoming microwaves would spur a cascade that would produce a flood of microwaves.

In December 1953 Townes and his students constructed a device that did exactly this, producing microwaves in a beam. They dubbed the process "microwave amplification by stimulated emission of radiation," which led to the more commonly used term *maser*. The maser quickly found many applications for its ability to send strong microwaves in any direction. The new device resulted in improvements in radar and also provided the basis for an atomic clock that was far more accurate than any mechanical timepiece ever invented. In the late 1950s Townes and his associates improved upon the maser by creating solid-state masers that could amplify ultraweak signals better than any other known means of amplification. In 1960 Townes developed the concepts for the visible-light maser, or laser (derived from "light amplification by stimulated emission of radiation"), which delivers infrared or visible light instead of microwaves. Two years later, American physicist Theodore H. Maiman built the first laser. Throughout the rest of his career, Townes's primary interest remained quantum theory, but he also pursued research in radio and infrared astronomy.

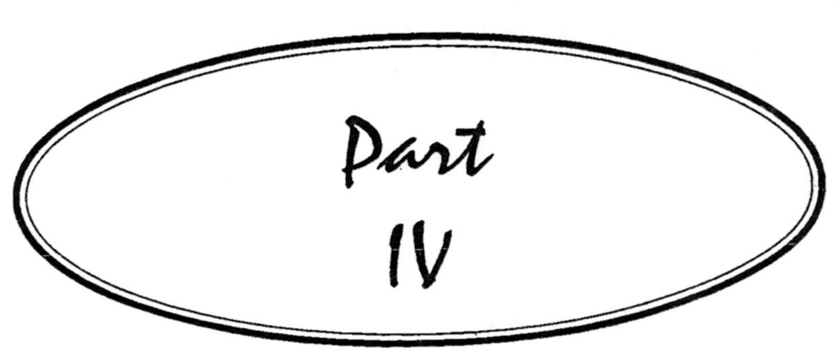

# Part IV

*1965* *Richard P. Feynman (United States)*
*Julian S. Schwinger (United States)*
*Sin-Itiro Tomonaga (Japan)*
*1966* *Alfred Kastler (France)*
*1967* *Hans A. Bethe (United States)*
*1968* *Luis W. Alvarez (United States)*
*1969* *Murray Gell-Mann (United States)*
*1970* *Louis Eugène F. Néel (France)*
*Hannes O.G. Alfvén (Sweden)*
*1971* *Dennis Gabor (United Kingdom)*
*1972* *John Bardeen (United States)*
*Leon N. Cooper (United States)*
*John R. Schrieffer (United States)*
*1973* *Ivar Giaever (United States)*
*Leo Esaki (Japan)*
*Brian D. Josephson (United Kingdom)*
*1974* *Sir Martin Ryle (United Kingdom)*
*Antony Hewish (United Kingdom)*
*1975* *Aage N. Bohr (Denmark)*
*Ben R. Mottelson (Denmark)*
*James Rainwater (United States)*
*1976* *Burton Richter (United States)*
*Samuel C.C. Ting (United States)*
*1977* *John H. Van Vleck (United States)*
*Philip W. Anderson (United States)*
*Sir Nevill F. Mott (United Kingdom)*
*1978* *Arno A. Penzias (United States)*
*Robert W. Wilson (United States)*
*Pyotr Leonidovich Kapitza (Union of Soviet Socialist Republics)*
*1979* *Steven Weinberg (United States)*
*Sheldon L. Glashow (United States)*
*Abdus Salam (Pakistan)*
*1980* *James W. Cronin (United States)*
*Val L. Fitch (United States)*
*1981* *Nicolaas Bloembergen (United States)*
*Arthur Leonard Schawlow (United States)*
*Kai Siegbahn (Sweden)*
*1982* *Kenneth G. Wilson (United States)*
*1983* *Subrahmanyan Chandrasekhar (United States)*
*William A. Fowler (United States)*
*1984* *Carlo Rubbia (Italy)*
*Simon van der Meer (Netherlands)*

Richard Feynman

*"American physicist Richard Feynman was well known for both his contributions to quantum electrodynamics and his enthusiastic teaching methods. Feynman reformulated quantum electrodynamic theory, which concerns the interactions between electromagnetic waves and matter. He is pictured here after winning the 1965 Nobel Prize in physics, which he shared with American physicist Julian S. Schwinger and Japanese physicist Sin-Itiro Tomonaga."*

# RICHARD PHILLIPS FEYNMAN

**Feynman, Richard Phillips** (1918-1988), American physicist and Nobel laureate. Feynman shared the 1965 Nobel Prize in physics for his role in the development of the theory of quantum electrodynamics, the study of the interaction of light with atoms and their electrons. He also made important contributions to the theory of quarks (particles that make up elementary particles such as protons and electrons) and superfluidity (a state of matter in which a substance flows with no resistance). He created a method of mapping out interactions between elementary particles that became a standard way of representing particle interactions and is now known as Feynman diagrams. Feynman was a noted teacher, a notorious practical joker, and one of the most colorful characters in physics.

Feynman was born in New York City. As a child he was fascinated by mathematics and electronics and became known in his neighborhood as "the boy who fixes radios by thinking." He graduated with a bachelor's degree in physics from the Massachusetts Institute of Technology (MIT) in 1939 and obtained a Ph.D. degree in physics from Princeton University in 1942. His advisor was John Wheeler, and his thesis, "A Principle of Least Action in Quantum Mechanics," was typical of his use of basic principles to solve fundamental problems.

During World War II (1939-1945) Feynman worked at what would become Los Alamos National Laboratory in central New

Mexico, where the first nuclear weapons were being designed and tested. Feynman was in charge of a group responsible for problems involving large-scale computations (carried out by hand or with rudimentary calculators) to predict the behavior of neutrons in atomic explosions.

After the war Feynman moved to Cornell University, where German-born American physicist Hans Bethe was building an impressive school of theoretical physicists. Feynman continued developing his own approach to quantum electrodynamics (QED) at Cornell and then at the California Institute of Technology (Caltech), where he moved in 1950.

Feynman shared the 1965 Nobel Prize in physics with American physicist Julian Schwinger and Japanese physicist Shin'ichirô Tomonaga for his work on QED. Each of the three had independently developed methods for calculating the interaction between electrons, positrons (particles with the same mass as electrons but opposite in charge) and photons (packets of light energy). The three approaches were fundamentally the same, and QED remains the most accurate physical theory known. In Feynman's space-time approach, he represented physical processes with collections of diagrams showing how particles moved from one point in space and time to another. Feynman had rules for calculating the probability associated with each diagram, and he added the probabilities of all the diagrams to give the probability of the physical process itself.

Feynman wrote only 37 research papers in his career (a remarkably small number for such a prolific researcher), but many consider the two discoveries he made at Caltech, superfluidity and the prediction of quarks, were also worthy of the Nobel Prize. Feynman developed the theory of superfluidity (the flow of a liquid without resistance) in liquid helium in the early 1950s. Feynman worked on the weak interaction, the strong force, and the composition of neutrons and protons later in the 1950s. The weak interaction is the force that causes slow nuclear reactions such as beta decay (the emission of electrons or positrons by radioactive substances). Feynman studied the weak interaction with American physicist Murray Gell-Mann. The strong force is the short-range force that holds the nucleus of an atom together.

Feynman's studies of the weak interaction and the strong force led him to believe that the proton and neutron were composed of even smaller particles. Both particles are now known to be composed of quarks.

The written version of a series of undergraduate lectures given by Feynman at Caltech, The Feynman Lectures on Physics (three volumes with Robert Leighton and Matthew Sands, 1963), quickly became a standard reference in physics. At the front of the lectures Feynman is shown indulging in one of his favorite pastimes, playing the bongo drum. Painting was another hobby. In 1986 Feynman was appointed to the Rogers Commission, which investigated the Challenger disaster-the explosion aboard the space shuttle Challenger that killed seven astronauts in 1986. In front of television cameras, he demonstrated how the failure of a rubber O-ring seal, caused by the cold, was responsible for the disaster. Feynman wrote several popular collections of anecdotes about his life, including "Surely You're Joking Mr. Feynman" (with Ralph Leighton and Edward Hutchings, 1984) and *What do YOU Care What Other People Think?* (with Ralph Leighton, 1988).

Julian Schwinger

*"American physicist Julian Schwinger won the 1965 Nobel Prize in physics. He won the award for his part in the development of a theory of quantum electrodynamics."*

## JULIAN SEYMOUR SCHWINGER

**Schwinger, Julian Seymour** (1918-1994), American physicist and Nobel Prize winner. Schwinger helped formalize the field of quantum electrodynamics, a discipline that combines quantum theory and relativity theory, and has to do with the behavior of charged particles in electrical fields. Schwinger shared the 1965 Nobel Prize in physics with Japanese physicist Shin'ichirô Tomonaga and American physicist Richard Phillips Feynman.

Born in New York City, Schwinger was a child prodigy, entering City College of New York (CCNY) at the age of 14. He later transferred to Columbia University and received his B.S. degree in 1936 at the age of 18. He stayed on at Columbia to earn his Ph.D. degree in 1939. Schwinger enjoyed a long and diverse professional career. Upon graduation, he became a researcher at the University of California at Berkeley from 1939 to 1941, where he worked with physicist J. Robert Oppenheimer. Schwinger left that position to accept a professorship at Purdue University in Indiana, where he worked from 1941 to 1943. He performed radar research at the Massachusetts Institute of Technology (MIT) from 1943 to 1945. Schwinger joined the faculty of Harvard University in 1945; two years later, he was a full professor, one of the very few to achieve such a status at Harvard while still in their 20s. He remained at Harvard until 1972. In 1975 Schwinger became a professor at the University of California, Los Angeles (UCLA), where he remained until his retirement in 1988.

Before formulating the contribution that would bring him fame, Schwinger studied work that other physicists had done in areas relating to quantum electrodynamics. Earlier work by British physicist Paul Dirac applied quantum mechanics to an analysis of the electromagnetic field. Dirac predicted that particles such as the electron would have an infinite quantity of energy, which led to other predictions that contradicted experimental observations. Schwinger reworked Dirac's mathematics in the theory so that those infinite quantities no longer appeared. This adjustment made Dirac's theory consistent with observation, and permitted physicists to predict the magnetic and other properties of particles and radiation. Tomonaga and Feynman made similar adjustments in Dirac's theory, working independently of Schwinger and each other.

Schwinger also postulated the existence of two different neutrinos associated with the electron and the muon (a negatively charged particle heavier than an electron), and predicted the discovery that massive charged particles carry weak nuclear forces. Experimental physicists later confirmed both of these theories.

Sin-Itiro Tomonaga

*"Japanese physicist Sin-Itiro Tomonaga won the 1965 Nobel Prize in physics. His work explored the electrodynamic properties of atomic particles."*

## SIN-ITIRO TOMONAGA

**Tomonaga, Sin-Itiro** (1906-1979), Japanese physicist and Nobel Prize winner. Tomonaga took an important step for quantum electrodynamics by making mathematical predictions that were consistent with observed physical phenomena of the special theory of relativity. For this work, he shared the 1965 Nobel Prize in physics with American physicists Julian Seymour Schwinger and Richard Phillips Feynman.

Born in Tokyo, Tomonaga became interested in physics after learning about American physicist Albert Einstein's well-publicized visit to Japan in 1922 and reading a book on relativity theory. He earned his B.S. degree in physics from Kyôto Imperial University in 1929. During his final year at Kyôto, Tomonaga studied quantum mechanics without the help of a professor. After graduating, he stayed on at Kyôto as an unpaid research assistant. In 1932 Tomonaga joined fellow researcher Yoshio Nishina at the Institute for Physical and Chemical Research to do theoretical and experimental work in nuclear physics. Tomonaga studied in Germany from 1937 to 1939 with German physicist Werner Heisenberg at the University of Leipzig. He then returned to Japan to obtain his Ph.D. degree in physics (1939) from Tokyo Imperial University (later renamed Tokyo University), where he later became a professor, a position he held until 1970.

During World War II (1939-1945), Tomonaga performed military research for the Japanese Navy and worked on theories of quantum electrodynamics. Earlier work by British physicist Paul Dirac applied quantum mechanics to an analysis of the electromagnetic field. Dirac predicted that particles such as the electron could have an infinite quantity of energy, which led to other predictions that contradicted experimental observations. Tomonaga reworked Dirac's mathematics in the theory so that those infinite quantities no longer appeared. This adjustment made Dirac's theory consistent with observation, and permitted physicists to predict the magnetic and other properties of particles and radiation. The renormalized theory of quantum electrodynamics has proved to be amazingly accurate, even with increasingly sophisticated and sensitive experimental equipment. Schwinger and Feynman also modified Dirac's theory. All three physicists worked independently of each other, but Schwinger and Tomonaga approached the problem in much the same way, while Feynman's method was very different.

After performing his groundbreaking research in quantum electrodynamics, Tomonaga worked in the areas of quantum dynamics, the theory of neutrons, and electromagnetics. He also became increasingly involved in scientific administration, serving as a member, and later president, of the Science Council of Japan in 1951. From 1957 on, he was active in movements against the deployment of nuclear weapons.

Alfred Kastler

*"French physicist Alfred Kastler won the 1966 Nobel Prize in physics. Kastler developed methods of using light to excite atoms, enabling the study of how electrons in atoms change energy levels."*

# ALFRED KASTLER

**Kastler, Alfred** (1902-1984), French physicist and Nobel Prize winner. Kastler developed methods that used light to manipulate and study the energy levels of electrons in atoms. For this work, he was awarded the 1966 Nobel Prize in physics.

Kastler was born in the town of Guebwiller in Alsace, which belonged to Germany until the 1919 Treaty of Versailles was signed after World War I (1914-1918), and which is now part of France. He earned a B.S. degree in 1926 at the prestigious École Normale Supérieure in Paris. After graduation, Kastler taught high school in several towns before accepting a position in 1931 as a research assistant at the University of Bordeaux, where he received his Ph.D. degree in physics in 1936. Kastler then became a lecturer at Clermont-Ferrand University from 1936 to 1938. He spent 1938 to 1941 as a professor of physics at the University of Bordeaux. In 1941 he returned to the École Normale Supérieure as assistant professor of physics; he was promoted to full professorship in 1945. Kastler retired in 1968, then served as director of research at the French National Center for Scientific Research from 1968 to 1972.

Kastler discovered and developed two methods-double resonance and optical pumping-for using light to energize atoms so that they can be studied. Both of these methods were great improvements on earlier methods and allowed much more detailed studies of atomic structure.

Double resonance, developed in the late 1940s, applies a beam of light and radio waves to a group of atoms. The frequency of the beam of light is carefully selected to excite atoms to a particular energy level. When the atoms return to their unexcited state, they emit energy in the form of light, which has characteristics specific to their energy level. If radio waves of just the right frequency are applied to the atoms after they have been excited by the beam of light, and before they return to their unexcited state, the characteristics of the atoms' energy level will change slightly. The light emitted by atoms in this state will be slightly different from that emitted by the atoms to which only the beam of light was applied. By finding the exact frequency of the radio waves that changes the characteristics of an energy level, Kastler could determine exactly how the energy level had changed and could then precisely map each energy level for the first time.

Optical pumping, developed in 1950, is a way to get a group of atoms into a uniform energy level. Polarized light is focused on a group of atoms in their unexcited state. Some atoms jump to a higher energy level and some do not. When the energized atoms return to the unexcited state and the light is focused on them again, none of the atoms will jump to a higher energy level, indicating that the energy states of all the atoms have the same characteristics. The technique of optical pumping was later applied in the development of the laser, the magnetometer (a device used to detect the presence of a metallic object), and improvements to the atomic clock.

**Hans Albrecht Bethe**

*"American physicist Hans Albrecht Bethe won the Nobel Prize in physics in 1967. He studied thermonuclear fusion, the process by which hydrogen is converted into helium."*

# HANS ALBRECHT BETHE

**Bethe, Hans Albrecht** (1906-), German-born American physicist and Nobel laureate, noted for his contributions to theories of stellar energy production. Bethe was born in Strasbourg, Alsace-Lorraine (then a part of Germany), and educated at the University of Frankfurt and the University of Munich, from which he received a Ph.D. degree in 1928. Bethe taught physics at various universities in Germany from 1928 to 1933 and in England from 1933 until 1935, when he began his long association with Cornell University.

Beginning in 1943 he worked at Los Alamos, New Mexico, on the atomic-bomb project. After initial misgivings he took part in the later development of the hydrogen bomb. At the same time he continued his work for the peaceful use and international control of nuclear energy. A prime advocate of the partial test-ban agreement signed by the United States, the Soviet Union, and the United Kingdom in 1963, he became an opponent of the Strategic Defense Initiative, proposed by the United States in the 1980s.

Bethe was awarded the 1967 Nobel Prize in physics for his studies of the production of energy by the sun and other stars, which he postulated occurs through thermonuclear fusion, a long series of nuclear reactions by which hydrogen is converted into helium. He was naturalized a U.S. citizen in 1941.

Luis Walter Alvarez

*"American Scientist Luis Walter Alvarez won the 1968 Nobel Prize in physics. Alvarez developed the liquid hydrogen bubble chamber, which he used to find atomic particles."*

# LUIS WALTER ALVAREZ

**Alvarez, Luis Walter** (1911-1988), American scientist born in San Francisco and educated at the University of Chicago, who won the 1968 Nobel Prize in physics for developing the liquid hydrogen bubble-chamber, with which he found atomic particles produced by high-energy nuclear events. He also developed the proton linear accelerator known as LINEAC. Alvarez had wide-ranging interests in science. In 1981 he and his son Walter, after studying geological strata, published a controversial theory that a giant meteorite striking the earth had caused the extinction of the dinosaurs.

**Murray Gell-Mann**

*"American physicist Murray Gell-Mann won the 1969 Nobel Prize in physics. He researched the interactions of elementary particles and advanced the quark theory."*

# MURRAY GELL-MANN

**Gell-Mann, Murray** (1929-), United States physicist, noted for his classification of subatomic particles and his proposal of the existence of quarks. Born in New York City, Gell-Mann attended Yale University and received a Ph.D. degree from the Massachusetts Institute of Technology in 1951. He taught at the University of Chicago from 1952 to 1955, when he joined the faculty of the California Institute of Technology. Gell-Mann was awarded the 1969 Nobel Prize in physics for work begun in Chicago in 1953. His research in particle physics concerned the interactions between protons and neutrons. On the basis of a proposed property called "strangeness," conserved by particles involved in strong and electromagnetic interactions ,Gell-Mann grouped related particles into multiplets, or families. In 1963 he and, independently, his colleague George Zweig advanced the quark theory; they hypothesized that quarks-particles carrying fractional electric charges-are the smallest particles of matter. Research in particle physics has made use of and thus far supported these theories.

Louis E. F. Neel

*"French physicist Louis E. F. Néel won the 1970 Nobel Prize in physics. He conducted ground breaking research on the magnetic properties of solids."*

# Louis Eugène Félix Néel

**Néel, Louis Eugène Félix** (1904-2000), French physicist and Nobel Prize winner. Néel made important advances in understanding the magnetic properties of solid materials, for which he shared the 1970 Nobel Prize in physics with Swedish physicist Hannes Olof Gösta Alfvén.

Born in Lyons, Néel completed his undergraduate studies at the Ecole Normale Supérieure in 1928. He went on to earn his Ph.D. degree from the University of Strasbourg in France in 1932. In 1937 he became a professor at the University of Strasbourg, where he stayed until 1945. During World War II (1939-1945), Néel also worked for the French Navy, developing ways to protect warships from magnetic mines. From 1945 until his retirement in 1976 he held various senior positions at the University of Grenoble.

In the late 1920s, when Néel began studying magnetism, physicists recognized only three types of magnetism: diamagnetism, paramagnetism, and ferromagnetism. Diamagnetic materials are slightly repelled by an external magnetic field, and paramagnetic materials are slightly attracted by an external magnetic field. Atoms in both of these types of substances tend not to orient themselves with each other, so the magnetic effects are almost completely canceled. In ferromagnetic materials, atoms tend to orient themselves in the same direction, so these materials have strong magnetic characteristics of their own.

Néel identified two additional types of magnetic materials that were added to the classification scheme: antiferromagnetic and ferrimagnetic. Antiferromagnetic substances act like paramagnetic substances above a certain temperature, called the *Néel point*. Below the Néel point, though, every atom oriented in one direction has a partner atom oriented in the opposite direction, so all magnetic effects are canceled. In ferrimagnetic substances, there are some pairs of atoms that cancel each other, but there is a net magnetic effect. For example, in the ferrimagnetic material magnetite, there are two atoms oriented in the same direction for every atom oriented in the opposite direction.

Néel's pioneering research on magnetics has provided new materials for the study of microwave electronics, computer memory, and other applications. In addition, Néel studied the "magnetic memory" of certain mineral deposits to learn of changes in the earth's magnetic field.

Hannes Alfvén

*"Swedish physicist Hannes Alfvén won the Nobel Prize in physics in 1970. Alfvén worked in magnetohydrodynamics (The study of plasmas in magnetic fields) and founded the field of plasma physics."*

# Hannes Olof Gösta Alfvén

**Alfvén, Hannes Olof Gösta** (1908-1995), Swedish physicist and Nobel laureate. For his discoveries in the field of plasma physics-the study of gaslike mixtures consisting of electrically charged particles found primarily in outer space-Alfvén was awarded the 1970 Nobel Prize for physics, which he shared with French physicist Louis Néel. Researchers have applied Alfvén's ideas to the study of sunspots, cosmic rays, and the origin of galaxies and the solar system. His work also has helped researchers develop thermonuclear reactors, devices that produce nuclear power.

Born in Norrköping, Sweden, Alfvén attended the University of Uppsala in Sweden, receiving a Ph.D. degree in 1934. Soon after graduating, he accepted a professorship at the University of Uppsala, where he remained until 1937. He went on to serve as a researcher at the Nobel Institute of Physics in Stockholm, Sweden, until 1940. After teaching abroad for several years, Alfvén became a professor at the Royal Institute of Technology in Stockholm. In 1967 he moved to the United States to teach at the University of California at San Diego.

Alfvén's research in plasma physics made him one of the founders of the field. He showed that a plasma has an *electric current* (a flow of charged particles) that produces a magnetic field. He also showed that, under certain conditions, the plasma *binds*, or freezes, the magnetic field, meaning that the plasma and the magnetic field move together. Physicists call this the frozen-in-flux theorem.

In 1939 Alfvén published a theory relating magnetic storms to the aurora. Magnetic storms occur when plasma streams from the sun enter the earth's upper atmosphere. The collisions between the energetic charged particles of the incoming plasma and the neutral gas molecules in the atmosphere release energy that is then seen as the light in the aurora. The aurora-commonly referred to as the aurora borealis (northern lights) or the aurora australis (southern lights) according to its location-occurs in high latitudes of both of the earth's hemispheres and consists of immense, rapidly shifting curtains and columns of pastel-colored lights. In his theory, Alfvén introduced a mathematical approximation that physicists now widely use to calculate the complex motion of a charged particle in a magnetic field.

Alfvén's work on the motion of electrically conducting fluids in a magnetic field was concerned primarily with geophysics and astrophysics. His research led him to postulate *"Alfvén's waves,"* transverse electromagnetic waves transmitted by plasma. Scientists later confirmed the waves in plasmas and liquid metals.

Dennis Gabor

*"British physicist Dennis Gabor won the 1971 Nobel Prize in physics. He developed the three-dimensional system of photography called holography."*

# DENNIS GABOR

**Gabor, Dennis** (1900-1979), British physicist, engineer, and Nobel laureate, born in Budapest, Hungary, and educated at the Technical University, Budapest, and the Technische Hochschule, Charlottenburg, Germany. He taught and performed research in Germany until 1933, when he immigrated to Great Britain. A naturalized British citizen, in 1949 he joined the faculty of the University of London.

Gabor concentrated his research on electron and plasma physics, electron microscopy, and physical optics. He became prominent for his invention (1947) and the subsequent development of holography, a lensless system of three-dimensional photography. For this work, Gabor was awarded the 1971 Nobel Prize in physics. His publications include *The Electron Microscope (1946) and Inventing the Future (1963).*

John Bardeen

*"American physicist John Bardeen won the Nobel Prize twice in physics one in 1956 and other in 1972. Bardeen helped developed the transistor an electronic device that performed many of the functions of the vacuum tube."*

## John Bardeen

**Bardeen, John** (1908-1991), American physicist and Nobel laureate, born in Madison Madison, Wisconsin, and educated at the University of Wisconsin and Princeton University. As a research physicist (1945-1951) at the Bell Telephone Laboratories in Murray Hill, New Jersey, he was a member of the team that developed the transistor, a tiny electronic device capable of performing most of the functions of the vacuum tube. For this work, he shared the 1956 Nobel Prize in physics with two colleagues, the American physicists William Shockley and Walter H. Brattain. Meanwhile he had joined (1951) the faculty of the University of Illinois. In 1972 he shared the Nobel Prize in physics with the American physicists Leon N. Cooper and John R. Schrieffer for the development of a theory to explain superconductivity, the disappearance of electrical resistance in certain metals and alloys at temperatures near absolute zero. Bardeen thus became the first scientist to win two Nobel Prizes in the same category.

Leon Cooper

*"American physicist Leon Cooper won the Nobel Prize in physics in 1972. Cooper developed a theory of why some metals can be superconductive."*

# LEON N. COOPER

**Cooper, Leon N.** (1930- ), American physicist, professor, and Nobel Prize winner. Cooper contributed significantly to the development of a theory of superconductivity by discovering what are now called "Cooper pairs"-that is, two electrons that, when situated a certain way among positive ions, no longer repel each other but instead develop an attraction for each other. These pairs then accumulate and move in the same direction; the result is a superconducting metal because there is no resistance to the flow of electricity through the metal. For his work in superconductivity, Cooper shared the 1972 Nobel Prize in physics with fellow American scientists John Bardeen and J. Robert Schrieffer.

Cooper was born in New York City. As a high school senior, he entered the Westinghouse Science Talent Search competition with a research project that analyzed how certain bacteria can become penicillin resistant. For his efforts, he was selected as one of 40 national winners. In 1954 Cooper received a Ph.D. degree in physics from Columbia University. His dissertation was supervised by American physicist J. Robert Oppenheimer, who had directed the development of the atomic bomb.

In 1955 Cooper was working on quantum field theory at the Institute for Advanced Study in Princeton, New Jersey, when John Bardeen, a scientist at the University of Illinois, invited Cooper to study superconductivity with him. First discovered in 1911, superconductivity is a phenomenon in which certain metals, when cooled to a temperature that is near absolute zero, no longer resist a flow of electricity running through them. The resulting free flow of electrons in the metal results in increased conductivity-an important discovery, particularly for the electronics industry. Electrical devices can waste large amounts of energy simply in trying to overcome electrical resistance, a problem that could be virtually eliminated if superconducting materials were used instead.

But for years after the discovery of superconductivity, scientists encountered difficulty when trying to apply the phenomenon in practical ways because of the extremely low temperature to which the metals must be cooled in order to overcome the electrical resistance. From 1955 to 1957 Bardeen, Cooper, and another physicist, J. Robert Schrieffer, developed what would later become known as the BCS theory of superconductivity, which explains why certain materials can be superconductive.

In 1957 Cooper left the University of Illinois to become an assistant professor of physics at Ohio State University. Since 1958 he has taught physics at Brown University in Providence, Rhode Island. Cooper holds honorary degrees from several United States universities and has received many awards for his work. In recent years, he has devoted his efforts toward a better understanding of memory and other brain functions.

John Schrieffer

*"American physicist John Schrieffer won the 1972 Nobel Prize in physics. He was awarded the prize for developing the first general theory of superconductivity."*

## JOHN ROBERT SCHRIEFFER

**Schrieffer, John Robert** (1931- ), American physicist and Nobel Prize winner. Schrieffer shared the 1972 Nobel Prize in physics with American physicists Leon N. Cooper and John Bardeen for the first general explanation of superconductivity, a complete lack of electrical resistance in a substance.

Born in Oak Park, Illinois, Schrieffer received his B.S. degree in physics from the Massachusetts Institute of Technology (MIT) in 1953. He earned his M.S. degree in 1954 and Ph.D. degree in 1957 from the University of Illinois. After he received his doctoral degree, he studied briefly in England and Denmark under a National Science Foundation (NSF) fellowship. Schrieffer taught physics at the University of Chicago from 1957 to 1960, at the University of Illinois from 1959 to 1960, and at the University of Pennsylvania in 1962. He taught at Cornell University from 1969 to 1975, and was a professor at the University of California at Santa Barbara from 1975 to 1992. Since 1992, Schrieffer has been a professor at Florida State University.

To explain superconductivity, Bardeen, Cooper, and Schrieffer had to explain how any substance could achieve perfect conductivity. Most metals lose resistance as they are cooled because the thermal vibrations of the atoms, which are a cause of resistance, slow down. But these vibrations will not cease completely until the substance is cooled to absolute zero (-273° C/-460° F), a temperature that cannot be obtained. Superconducting material reaches zero resistance suddenly and at temperatures above absolute zero.

In 1956 Cooper showed that electrons carrying current through a superconductor acted strangely-instead of repelling each other because of their like charge, pairs of electrons actually seemed to be bound together. At low enough temperatures, the atomic vibrations in the crystal structure of the metal allow one of the electrons to draw the positively charged metal atoms to itself. This creates a concentration of positive charge, which attracts the second electron. Such a pair of electrons is now called a Cooper pair.

At the University of Illinois in 1956 and 1957 Schrieffer and Bardeen sought to apply this model of electron behavior to a large percentage of the electrons in a superconductor. They collaborated with Cooper, also at the University of Illinois, to show that the interaction between Cooper pairs forces many of the free electrons to march along in formation. Below a certain temperature, the forces holding the pairs of electrons together are stronger than the thermal vibration of the atoms. Without anything to slow down the electrons, there is no resistance.

This theory of superconductivity is known as the Bardeen-Cooper-Schrieffer (BCS) theory. It has led to the development of superconductors that can operate at higher temperatures and that can be used to make strong electromagnets that use little power. Such magnets have many potential uses, including applications in quantum mechanics, biology, generation of electricity, and many other fields.

In addition to the Nobel Prize, Schrieffer has received many honorary degrees and other awards. He became the president of the American Physical Society in 1996.

Ivar Giaever

*"Norwegian-born American physicist Ivar Giaever won the Nobel Prize in physics in 1973. Giaever researched semiconductors and superconductors and discovered that superconductivity works better at very low temperatures."*

# IVAR GIAEVER

**Giaever, Ivar** (1929- ), Norwegian-born American physicist who shared the 1973 Nobel Prize in physics with Japanese physicist Leo Esaki and Welsh physicist Brian David Josephson for his work on "tunneling" effects in semiconductors and superconductors. According to classical physics, an electric current cannot flow in a circuit interrupted by an insulating barrier—that is, when electrons reach the "wall" of insulating material, they cannot continue forward. Since the 1930s, quantum theory had predicted that electrons might be able to "tunnel" through an insulating barrier if it were thin enough. Giaever discovered that cooling two semiconductors to near-absolute zero temperatures greatly enhances superconductivity. As a result of these studies, he developed a technique that allowed physicists to easily observe and measure semiconductor properties. This technique allowed Giaever to confirm certain previously unproven theories about superconductivity.

Giaever was born in Bergen, Norway. He worked at a munitions plant for a year after high school before enrolling in the Norwegian Institute of Technology. He earned a B.S. degree there in mechanical engineering in 1952. He served in the Norwegian Army from 1952 to 1953, and then accepted his first professional position as a patent examiner for the Norwegian government. In 1954 Giaever moved to Canada, taking a job as a mechanical engineer with the Advanced Engineering Program of the General Electric Company. General Electric sent him to an advanced engineering program in its Schenectady, New York, offices in 1956. He remained in New York, becoming an applied mathematician at the company's research and development center. In 1958 he became interested in physics, taking a position in a solid-state physics research group at General Electric. He then returned to school to work toward a Ph.D. degree in physics at night. In 1964 he received his doctorate from Rensselaer Polytechnic Institute. He also became a U.S. citizen that year. Shortly thereafter, Giaever began his research on tunneling effects in semiconductors and superconductors.

During the 1970s, Giaever's work increasingly began to focus on physics and its relationship to biology. He temporarily left General Electric in 1970 to study biophysics at the University of Cambridge on a Guggenheim Fellowship. When he returned to General Electric, he began investigating the properties of cell membranes and how proteins interact with solid surfaces. He has also studied stress reactions and antibody-antigen reactions. This work has been conducted at both General Electric and the Albany Medical Center in Albany, New York. Giaever spent a year as an adjunct professor at the University of California, San Diego and as a visiting professor at the Salk Institute for Biological Studies in La Jolla, California in 1975. The Rensselaer Polytechnic Institute made him Institute Professor of Science in 1988 and the University of Oslo in Norway appointed him professor-at-large that same year. In addition to his academic activities, Giaever is president of applied Biophysics Inc., a company started in 1991 that develops applications for electric sensor devices

Leo Esaki

*"Japanese physicist Leo Esaki won the 1973 Nobel Prize in physics. He proved the concept of tunnelling in semiconductor and developed the tunnel diode."*

# LEO ESAKI

**Esaki, Leo** *(1925- ),* Japanese physicist whose groundbreaking work on semiconductors earned him a Nobel Prize in 1973. Esaki proved the concept of tunnelling in semiconductors and developed the tunnel diode, also known as the Esaki diode. In classical physics, an electric current cannot flow in a circuit interrupted by an insulating barrier-that is, when electrons reach the "wall" of insulating material, they cannot continue forward. Since the 1930s, quantum mechanics, the branch of physics that studies the motion of subatomic particles and related phenomena, had predicted that electrons might be able to "tunnel" through an insulating barrier if it were thin enough. Esaki developed a diode with electrical junctions only 10 billionths of a meter thick through which electrons could tunnel. Esaki shared the Nobel Prize with Norwegian-born American physicist Ivar Giaever and American physicist Brian D. Josephson.

Esaki was born in Ôsaka, Japan. He attended the University of Tokyo, where he earned a B.S. degree (1946), an M.S. degree (1947), and a Ph.D. degree (1959) in physics. While studying for his doctorate, he worked in Tokyo at Kôbe Kogyo Corporation and Sony Corporation. His discovery of tunneling occurred at Sony in 1957 while he was still a student. Esaki moved to the United States in 1960 to join the Thomas J. Watson Research Center at IBM in New York. His research focused on semiconductor physics. He was made an IBM fellow, the company's highest research position, in 1965. He began work in superlattices as part of an effort to demonstrate other predicted but unproven theories of quantum mechanics. *Superlattices* are synthetic crystals composed of extremely fine layers of different semiconductors. One of the potential uses for this material is in high-speed computers. Esaki stayed at IBM for 33 years, eventually becoming a director of the company. During this period, he also lectured at the University of Pennsylvania and the University of Tokyo. When he retired from IBM in 1993, Esaki returned to Japan. Since then he has served as president of Tsukuba University.

Brian David Josephson

*"Welsh physicist Brian David Josephson won the 1973 Nobel prize in physics. Josephson's theories predicted how electrons flow through an insulating barrier, a phenomenon now called the Josephson effect."*

# Brian David Josephson

**Josephson, Brian David** (1940- ), Welsh physicist and one of the youngest recipients of a Nobel Prize. Josephson shared the 1973 Nobel for physics with Japanese physicist Leo Esaki and Norwegian-born American physicist Ivar Giaever for their research on tunnelling effects in semiconductors and superconductors. According to classical physics, an electric current-and specifically, electrons-cannot flow in a circuit that is interrupted by an insulating barrier. Since the 1930s, many physicists had predicted that electrons might be able to "tunnel" through an insulating barrier if it was thin enough. Esaki, who first demonstrated tunnelling in semiconductors, laid the groundwork for Giaever's research on superconducting tunnel junctions. Based on their discoveries, Josephson formulated theories predicting how electrons flow through a tunnel barrier.

Josephson was born in Cardiff, Wales. He attended Trinity College at the University of Cambridge, where he earned a B.A. (1960), an M.A. (1964), and a Ph.D. (1964) in physics. In 1962 he made his Nobel-prize winning observations about the behavior of an electrical contact between a superconducting material and a normal metal separated by a very thin insulating layer. Traditional quantum theory stated that only a small amount of current (electrons) could tunnel through the nonconducting barrier. Josephson predicted that a much higher number of electrons would actually move across the insulator. He also noted that this current would be affected by an external magnetic field. The flow of electric current through nonconductive material became known as the Josephson effect. Josephson's discoveries have had practical applications in the development of miniature electronics.

In 1964 Josephson took a teaching position at the University of Cambridge. He spent a year at the University of Illinois as a visiting research professor, returning to the University of Cambridge in 1967 to serve as assistant director of research in physics. He then worked as reader in physics for two years. In 1974 Josephson became professor of physics.

He shifted his research focus from physics to the scientific study of the mind after attending a 1971 lecture on transcendental meditation. He became interested in synthesizing modern physics and mathematics with the study of intelligence, language, higher states of consciousness, and the paranormal. With Indian chemical engineer V.S. Ramachandran, Josephson edited *Consciousness and the Physical World* (1979).

Sir Martin Ryle

*"British astronomer Sir Martin Ryle won the 1974 Nobel Prize in physics. He won the award for his development of aperture synthesis in radio astronomy, whereby two radio telescopes are operated at once to gather data on distant objects."*

# SIR MARTIN RYLE

**Ryle, Sir Martin** (1918-1984), British astronomer and Nobel laureate, known for his developmental work in radio astronomy. Ryle was born in Brighton, Sussex, and educated at the University of Oxford. After World War II he held a fellowship at Cavendish Laboratory, University of Cambridge. A lecturer, then professor of radio astronomy at Cambridge, he served from 1957 as director of the university's Mullard Radio Astronomy Observatory. Ryle made studies of celestial radio sources and directed the compilation of the Cambridge catalogs of radio sources, which provided the basis for the discovery of quasars. In 1966 he was knighted, and from 1972 to 1982 he was Britain's Astronomer Royal. In 1974 Ryle was awarded the Nobel Prize in physics for his development of aperture synthesis in radio astronomy. He shared the prize with the English physicist Antony Hewish.

Antony Hewish

*"British astronomer Antony Hewish won the 1974 Nobel Prize in physics. Hewish discovered the celestial objects now known as pulsars."*

# ANTONY HEWISH

**Hewish, Antony** (1924–), British astronomer and Nobel Laureate who in 1967 with assistant Jcelyn Bell discovered the celestial objects now called pulsars (see star). He was born in Fowey England and educated at the university of Cambridge. For their separate work he and British astronomer Sir Martin Ryle were awarded the Nobel Prize in physics in 1974.

Aage Niels Bohr

*"Danish physicist Aage Niels Bohr won the Nobel Prize in physics in 1975. He developed the collective-motion theory of the atomic nucleus that helped explain many nuclear properties."*

# Aage Niels Bohr

**Bohr, Aage Niels** (1922-), Danish physicist and Nobel laureate, born in Copenhagen. The son of Niels Bohr, he assisted his father on the atomic bomb project at Los Alamos, New Mexico, during World War II. He then joined the Institute for Theoretical Physics in Copenhagen, devoting his attention to the inner structure of the atom.

In 1954 he wrote his doctoral thesis at the University of Copenhagen. It dealt with a collective motion theory of the atomic nucleus that he had developed with the U.S. physicist Ben R. Mottelson at the suggestion of the U.S. physicist James Rainwater. The theory helped to explain many nuclear properties by showing that nuclear particles can vibrate and rotate so as to distort the shape of the nucleus from the expected spherical symmetry into an ellipsoid. Bohr, Mottelson, and Rainwater received the 1975 Nobel Prize in physics for this work.

In 1963 Bohr became director of the institute, now renamed in honor of his father. He resigned in 1970 to devote more time to research, but in 1975 he became director of the Nordic Institute of Theoretical Atomic Physics, which shares research and facilities with the Niels Bohr Institute.

Benjamin R. Mottelson

*"Danish physicist Benjamin R. Mottelson won the 1975 Nobel Prize in physics. Mottelson investigated the structure of the atomic nucleus and helped develop a new model that combined previous concepts of the nucleus."*

# BENJAMIN ROY MOTTELSON

**Mottelson, Benjamin Roy** (1926- ), American-born Danish physicist whose research on the nucleus of atoms resulted in his winning the 1975 Nobel Prize in physics with two colleagues, the Danish physicist Aage Niels Bohr and the American physicist Leo James Rainwater. The three men received the award for developing a collective model of the structure of the nucleus of an atom. They combined concepts proposed in two earlier models: the liquid drop model, which described nucleus behavior as liquidlike; and the shell model, which explained orbital behavior of nucleons within the nucleus. The Nobel committee's citation noted that the new model made the "connection between collective motion and particle motion in atomic nuclei and the development of the theory of the structure of the atomic nucleus based on this connection."

A Chicago native, Mottelson spent time in the United States Navy during World War II (1939-1945). As part of his military service, he was sent to officer training school at Purdue University in West Layfayette, Indiana. He reenrolled at Purdue at the end of the war, graduating in 1947 with a B.S. degree. He then attended Harvard University in Cambridge, Massachusetts, earning an M.A. degree in physics in 1948 and a Ph.D. degree in physics in 1950. His adviser for his doctoral thesis, on nuclear physics, was Nobel Prize-winning American physicist Julian Schwinger.

Mottelson's career in Denmark started in 1950 with a fellowship he received from Harvard to study at the Institute for Theoretical Physics in Copenhagen. He stayed at the Institute (later renamed Niels Bohr Institute) for several more years on additional fellowships from the U.S. Atomic Energy Commission (1951-1953) and the European Organization for Nuclear Research (Conseil Européen pour la Recherche Nucléaire or CERN) (1953-1957). He has conducted research at the Nordic Institute for Theoretical Atomic Physics (NORDITA) since 1957. Mottelson formed his Nobel Prize-winning alliance with Bohr at the Institute for Theoretical Physics and at NORDITA. In 1981 he replaced Bohr as director of NORDITA.

In 1957 Mottelson and Bohr authored *Collective and Individual Particle Aspects of Nuclear Structure.* In 1969 they published the first volume of *Nuclear Structure*, a three-volume set. That same year Mottelson received the Atoms for Peace Award of the Ford Motor Company Fund. In 1971 he became a Danish citizen. He has received honorary degrees from the University of Heidelberg and Purdue University

L. James Rainwater

*"American physicist L. James Rainwater won the 1975 Nobel Prize in physics. He conceived a new model of the atomic nucleus that resolved the problems of previous models."*

# LEO JAMES RAINWATER

**Rainwater, Leo James** (1917-1986), American physicist and Nobel laureate in physics. Rainwater shared the 1975 Nobel Prize for physics with Danish physicists Ben Mottelson and Aage N. Bohr. The three physicists won the prize for work unifying two theoretical models of the atomic nucleus.

Rainwater was born in Council, Idaho. He graduated from the California Institute of Technology in 1939 with a bachelor's degree in physics. He received his doctoral degree from Columbia University in 1946 for work he did at Columbia for the Manhattan Project, the group of scientists who developed the atomic bomb during World War II (1939-1945). Rainwater became a professor at Columbia in 1952, and remained there, teaching and doing experimental work in physics, until he retired in 1986. Rainwater also served as director of the university's Nevis Cyclotron Laboratory from 1951 to 1953 and from 1956 to 1961.

In 1949 Rainwater started studying the fundamental structure of the center and largest part of an atom that holds the neutrons and protons, called the *nucleus.* At that time, there were two basic ideas about how the nucleus behaved. Danish physicist Niels Bohr (the father of Aage Bohr, one of the scientists with whom Rainwater shared the Nobel Prize) had formulated the liquid-drop model for the nucleus in 1936. This model describes the nucleus as a drop of liquid, capable of changing shape. The shell model was developed by the German American physicist Maria Goeppert-Mayer and German physicist Hans Daniel Jensen in 1949. In this model, the particles (protons and neutrons) inside the nucleus (nucleons) move in concentric orbits around the center of the nucleus.

Both models had problems. The liquid-drop model helped explain nuclear fission, but could not explain other important nuclear phenomena. The shell model also offered explanations for some things and failed on others. Rainwater realized that he could solve the discrepancies produced by both models by thinking of the nucleus as oblong in some situations.

Rainwater justified his oblong nucleus by developing a model that drew from the shell model. He proposed that the outer orbit of particles in some nuclei did not have the right number of particles for the forces between the particles to balance each other and make the nucleus a sphere. Instead, it would be deformed by the forces of the particles in the inner orbits on the particles in the incomplete outer orbit. Mottelson and Aage Bohr confirmed Rainwater's theory in 1953, after using the model to calculate how a nucleus would behave in certain situations and comparing their results to experimental data.

Burton Richter

*"American physicist Burton Richter won the 1976 Nobel Prize in physics. Richter shared credit for the discovery of the subatomic particle J/psi."*

# Burton Richter

**Richter, Burton** (1931- ), American physicist and Nobel laureate. In 1976, Richter shared the Nobel Prize for physics with American physicist Samuel C.C. Ting for their independent discoveries of the subatomic particle J/psi.

Born in New York City, Richter received his bachelor's degree in 1952 and his doctoral degree in 1956 from the Massachusetts Institute of Technology (MIT). In 1956 he became a research associate at Stanford University. He was made an assistant professor at Stanford in 1960, and became an associate professor at the Stanford Linear Accelerator Center (SLAC) in 1963. In 1967, he was made a full professor of physics at Stanford. He has been the director of SLAC since 1984.

Richter's research focused on high-energy particle physics, the study of subatomic particles moving extremely fast. To explore this field, physicists have built particle accelerators, devices used to speed charged elementary particles to high energies. The accelerators Richter used at MIT were cyclotrons, which move particles in a spiral path as they gained speed, and synchrotrons, which move particles in a circular path. Both these accelerators direct the high-energy particles at stationary atoms when the particles have reached the desired speed. The linear accelerator at SLAC can generate positively charged particles as well as the negatively charged particles used in other accelerators. Because of the particles' opposite charges, the magnetic field used to contain and accelerate them affect them differently. The positively charged particles, or positrons, move through the accelerator counterclockwise, while the negatively charged particles, or electrons, move clockwise in the same area. Thus physicists can create much higher-energy collisions between the two moving beams of particles than they could with one moving beam of particles and stationary atoms.

In 1973, Richter began experimenting with a device, called the Stanford positron-electron accelerating ring (SPEAR), designed to create those high-energy collisions between positrons and electrons. When the positrons and electrons collided, they produced a burst of electromagnetic energy out of which other particles were produced. In 1974, Richter's group was studying the rate at which such collisions produced hadrons, a class of particles related to the proton and neutron. They noticed that the production rate of hadrons peaked sharply, often a sign of a new particle, at a particular energy. Richter gave the new particle the name psi. At the same time, Ting had discovered the particle using a different technique at MIT. Ting had named the particle *J*, so the two names were joined into J/psi. The discovery of the J/psi particle gave experimental evidence of the theorized fourth elementary particle known as a quark. This fourth quark was called charm.

Samuel C. C. Ting

*"American physicist Samuel C. C. Ting won the 1976 Nobel Prize in physics. Focusing on high-energy particle physics, he discovered a new elementary particle, J/psi."*

# Samuel Chao Chung Ting

**Ting, Samuel Chao Chung** (1936- ), American physicist and Nobel Prize winner. In 1976 Ting shared the Nobel Prize in physics with fellow American physicist Burton Richter for independently discovering the elementary particle J/psi.

Ting was born in Ann Arbor, Michigan, but spent most of his childhood in mainland China and Taiwan. After returning to the United States in 1956, Ting received his B.S. degree in mathematics and physics in 1959, his M.S. degree in 1960, and his Ph.D. degree in 1962, all from the University of Michigan. In 1963 he worked at the European Organization for Nuclear Research (CERN, now the European Laboratory for Particle Physics) in Switzerland. Two years later, he became assistant professor of physics at Columbia University. Ting did research in Hamburg, Germany, at the Deutches Elektronen-Synchrotron (DESY) facility in 1966. In 1967 he accepted a position at the Massachusetts Institute of Technology (MIT) and in 1969 became a full professor there. He currently continues his research at MIT.

Ting's research centers on high-energy particle physics, the area of physics that deals with the structure of elementary particles. One way physicists study these particles is by smashing particles together and looking at what comes out of the collision. That is accomplished with particle accelerators, devices designed to speed up charged particles to the desired velocity for the collision.

In 1971 Ting and a team of physicists began conducting particle research at the Brookhaven National Laboratory in Upton, New York. The group was searching for new particles, specifically relatively heavy particles with short lives. They used a proton accelerator that could hit a stationary target (Ting's group used beryllium as a target) with 10 trillion protons per second. Protons are heavier particles than the electrons and positrons used in some particle accelerators, and so are more likely to produce heavier particles in collisions.

In August of 1974 Ting's group discovered that collisions of a certain energy were producing a large number of electrons and positrons. Ting concluded that these particles were the result of the decay of an unknown, short-lived particle produced by collisions at that energy. He named this new particle J.

At a meeting at the Stanford Linear Accelerator Center (SLAC) in California, Ting learned of Richter's discovery of the new particle that Richter called *psi.* The two physicists compared findings, concluded they had found the same particle, and combined the names to call the particle J/psi.

The discovery of the J/psi particle provided evidence for the existence of a fourth quark, called charm, which had been predicted but never proven to exist. Quarks are the building blocks of elementary particles, and physicists now believe there are six types of quarks.

John Van Vleck

*"United States physicist John Van Vleck won the 1977 Nobel Prize in physics. Known as the "father of modern magnetism" Van vleck explored the development of semiconductors, materials whose electrical conductivity varies as the temperature changes."*

# JOHN HASBROUCK VAN VLECK

**Van Vleck, John Hasbrouck** (1899-1980), American physicist and Nobel Prize winner, known as "the father of modern magnetism." Van Vleck shared the 1977 Nobel Prize in physics with American physicist Philip Anderson and British physicist Sir Nevill Mott for their contributions to the study of semiconductors, materials with electrical resistance that varies with temperature and voltage.

Van Vleck was born in Middletown, Connecticut. He received his B.S. degree from the University of Wisconsin in 1920. He earned his M.S. and Ph.D. degrees from Harvard University in 1921 and 1922, respectively. Van Vleck spent one year as a physics instructor at Harvard, then taught the subject at the University of Minnesota from 1923 to 1928. He joined the physics department at the University of Wisconsin in 1928 and stayed until 1934, when he returned to Harvard. During World War II (1939-1945), Van Vleck studied the workability of nuclear weapons at the University of California at Berkeley and worked on improving radar technology at Harvard's Radio Research Laboratory. After the war, he served as the chairperson of the physics department and the dean of engineering and applied physics, and held the Hollis Chair in Mathematics and Natural Philosophy at Harvard. In 1969 he became professor emeritus.

Most of Van Vleck's work focused on the study of magnetism at a quantum-mechanical level . In 1932 he published *The Theory of Electrical and Magnetic Susceptibility*, which showed the quantum-mechanical explanations for many phenomena that occur in matter. The book became an important text in the developing field of solid-state physics.

Van Vleck applied quantum mechanics to ferromagnetism-the type of magnetism in ordinary magnets-and paramagnetism-the type of magnetism displayed by substances that have slight, fairly constant sensitivity to being magnetized. He also developed a theory that allowed scientists to calculate the energy levels of an atom or ion in a crystal, important to the development of lasers. In addition, he worked on magnetic resonance; magnetic resonance imaging (MRI) has become an important tool in medicine.

Van Vleck's fundamental work on magnetism complemented and served as a base to Anderson and Mott's work on disordered systems. Together, their research contributed to developments in the field of semiconductors.

Philip Anderson

*"American physicist Philip Anderson won the Nobel Prize in physics in 1977. Anderson helped develop theories of magnetism and conduction in solid-state physics."*

# PHILIP WARREN ANDERSON

**Anderson, Philip Warren** (1923- ), American physicist and Nobel laureate who helped develop basic theories of magnetism and conduction in solid-state physics. Anderson shared the 1977 Nobel Prize for physics with British physicist Sir Nevill Francis Mott and American physicist John Hasbrouck Van Vleck.

Anderson was born in Indianapolis, Indiana. He earned his bachelor's, master's and doctoral degrees from Harvard University, where he was a student of John Van Vleck. Anderson studied at Harvard University from 1939 to 1949, except for the years 1943 to 1945, when he interrupted his education to work as a radio engineer in the United States Navy during World War II.

After he earned his doctorate from Harvard, Anderson went to work at Bell Laboratories in New Jersey. Anderson worked for Bell Laboratories until his retirement in 1984, while serving terms as professor of physics at several universities, including the University of Tokyo and Princeton University. From 1967 to 1975 Anderson was a visiting professor of theoretical physics at the University of Cambridge in England, where he worked with Nevill Mott.

Anderson's work focused on solid-state physics, also known as condensed-matter physics. One of the major advances of solid-state physics was the discovery of semiconductors in the 1920s. A semiconductor is a material that conducts electricity at room temperature more readily than an insulator, but less easily than a metal. Use of semiconductors made possible the development of integrated circuits, the components that control many electronic devices such as computers. One of the most important semiconductors is the crystalline form of silicon. Anderson's work concentrated on solids with no crystalline structure, which are called *amorphous* materials.

Anderson's work with amorphous materials like amorphous silicon led him to demonstrate in 1958 that it is possible for an electron to get trapped in a small area. This phenomenon, known as "Anderson localization," suggests that amorphous materials can be used in place of the crystalline semiconductors used today. Anderson's discoveries also led to the development of electronic switching and memory devices made from amorphous materials such as glass. The field of amorphous semiconductors has become an area of intense research since Anderson's work.

Sir Nevill Mott

*"British physicist Sir Nevill Mott won the 1977 Nobel Prize in physics. His pioneering research helped clarify the electrical conductivity properties of noncrystalline and impure crystalline structures."*

# Sir Nevill Francis Mott

**Mott, Sir Nevill Francis** (1905-1996), British physicist and Nobel laureate known for his investigations of semiconductor physics. Mott shared the 1977 Nobel Prize in physics with American physicists Philip W. Anderson and John Hasbrouck Van Vleck for their work on the electronic properties of noncrystalline and impure crystalline structures.

Mott was born in Leeds, England, and earned a bachelor's degree in mathematics and physics from Saint John's College at the University of Cambridge, England, in 1927. Over the next three years he worked at a number of institutions in England, Denmark, and Germany. He received his master's degree from Cambridge in physics in 1930 and in 1933 Mott was appointed professor of theoretical physics at the University of Bristol. During World War II (1939-1945), Mott worked on a theory of decision making for strategic planning for the British military. He returned to Bristol after the war, and in 1948 was made professor of physics and director of a physics laboratory. In 1954 he became professor of physics at Cambridge, a position he held until his retirement in 1971. After retirement, he continued his research in noncrystalline materials. Mott was knighted in 1962.

While studying the properties of metallic elements, Mott became interested in refining the quantum mechanical band theory. According to this theory, electrons in a substance can only occupy certain bands of energies that are specific to that substance. In the late 1940s Mott added calculations to account for the interactions between electrons and explained why some materials can change from materials that carry electric current well, or conductors, to materials that carry electric current poorly, or *insulators*. These materials that can change conductivity are called *semiconductors*, and made possible the development of integrated circuits, the components that control many electronic devices. The changes in properties that define semiconductors are now called Mott transitions and were important in semiconductor development.

One of the most important semiconductors is the crystalline form of silicon. In the 1960s Mott became interested in a paper written by Anderson, which considered the semiconductor properties of materials without crystalline structure, known as *amorphous materials*. Anderson and Mott worked together at Cambridge from 1967 to 1975. By working with Anderson, Mott was able to explain why an insulator can be made to carry an electric current by adding impurities to the insulating material and why some amorphous materials will carry only no electric current or electric current above a certain value, a phenomenon known as *minimum conductivity*.

Arno Penzias

*"American astrophysicist Arno Penzias won the 1978 Nobel Prize in physics. He won the award for observations supporting the big-bang theory of how the universe was formed."*

## ARNO PENZIAS

**Penzias, Arno** (1933- ), German-born American radio engineer who, with American engineer Robert Wilson, was the first to detect the cosmic microwave background radiation. This radiation seems to be evenly distributed throughout the universe, leading scientists to believe that it is the cooling remains of energy released at the big bang, the explosion that released all the matter and energy in the universe. The background radiation represents some of the strongest evidence in favor of the big bang theory. Penzias and Wilson shared half of the 1978 Nobel Prize in physics for their discovery. The other half of the prize went to Soviet physicist Peter Kapitza for his work in low-temperature physics. See also Radio Astronomy.

Penzias was born in Munich, Germany. His parents fled Nazi Germany in 1940 and emigrated to the United States, taking their two young sons with them. Studying at the City College of the City University of New York, Penzias earned his bachelor's degree in physics in 1954. He served in the U.S. Army Signal Corps for two years, then returned to New York. He continued his studies at Columbia University, where he was awarded his master's degree in 1958 and his doctorate in 1962, both in physics.

Since 1961 Penzias has been associated with the Radio Research Laboratories at Bell Laboratories in New Jersey. When Penzias began work at Bell Labs, the laboratory was part of AT&T Corp. Since 1996 the laboratory has been a division of Lucent Technologies, a spin-off company of AT&T. From 1961 to 1972, Penzias was a staff member of the radio research department, and from 1972 to 1976, he was head of the radiophysics research department. In 1976 he became the director of the Radio Research Laboratory, and he served as director of the Communications Sciences Research Division from 1979 to 1981. From 1981 to 1995, Penzias was vice president of research for Bell Labs. He served as vice president and chief scientist from 1995 to 1998. Penzias retired from his vice presidency and chief scientist position in 1998 to work as an advisor and spokesperson for Lucent Technologies.

In addition to his posts in the telecommunications industry, Penzias also concurrently held a series of academic positions. The first of these was as lecturer in the department of astrophysical science at Princeton University in New Jersey from 1967 to 1982. After that, he held many yearlong honorary lecturer positions at diverse institutions, including the National Radio Astronomical Observatory in West Virginia and Stanford University in California.

In 1963 Penzias and Wilson were assigned to trace the source of radio noise that was interfering with the development of a communications program involving satellites. By May 1964 the two physicists had detected a surprisingly high level of radiation at a wavelength of 7.3 cm (2.9 in). The radiation seemed to have no particular source; it was *isotropic*, or came equally from all directions. Penzias and Wilson excluded all known terrestrial sources of such radiation and still found that the noise they were detecting was one hundred times more powerful than they could account for.

Penzias and Wilson took this result to Robert Dicke, professor of physics at Princeton University. Dicke was interested in microwave radiation and had predicted that this sort of radiation should be present in the universe as a residue of the intense heat associated with the birth of the universe following the big bang. His department was in the process of constructing a radio telescope designed to detect precisely this radiation, at a wavelength of 3.2 cm (1.3 in), when Penzias and Wilson presented their data.

Since Penzias and Wilson's detection of this radiation, background radiation has been the subject of intense study. The wavelengths at which it exists and its peak intensity very closely match the pattern of radiation that a blackbody, or a perfectly radiating object, would emit at a certain temperature. The temperature that the background radiation seems to represent is just below -270° C (just below -454° F). For cosmologists, the background radiation is the most convincing evidence in favor of the big bang model for the origin of the universe.

Penzias's later work has been concerned with developments in radio astronomy, instrumentation, satellite communications, and atmospheric physics.

Robert W. Wilson

*"American physicist and radio astronomer Robert W. Wilson won the 1978 Nobel Prize in physics. While attempting to measure the intensity of radiation from a single, specific point in the sky, he discovered cosmic microwave background radiation."*

# ROBERT WOODROW WILSON

**Wilson, Robert Woodrow** (1936- ), American physicist, radio astronomer, and Nobel Prize winner. Wilson shared the 1978 Nobel Prize in physics with German-born American physicist Arno A. Penzias for their 1965 detection of long-wavelength, low-temperature radio emissions coming from the entire universe, and with Soviet physicist Peter Kapitza, for his studies of liquid helium.

Born in Houston, Texas, Wilson received his B.S. degree in physics from Rice University in 1957. In 1962 he received his Ph.D. degree from the California Institute of Technology (Caltech), and that same year became a research fellow at Caltech. In 1963 he went to work at the Radio Research Laboratory of Bell Telephone Laboratories in New Jersey. In 1976 he was appointed head of the Radio Physics Research Department at Bell Labs. Wilson became an adjunct professor at the State University of New York in 1978. Today he continues his research at Bell Labs.

At Bell Labs, Wilson and Penzias worked with a 6-m (20-ft) antenna converted into a radio telescope. The precision of the instrument and the scientists' ability to account for radio waves from the ground, the atmosphere, and other local sources enabled Wilson and Penzias to measure the intensity of the radiation from a point of interest in the sky.

In 1964 the two scientists used their system for the first time, measuring radio waves from the remnants of a supernova in the constellation Cassiopeia. Despite all their efforts to account for every source of radiation that could interfere with their measurements, the researchers found they were picking up background noise. They tried measurements in many different areas of the sky, but the background radiation was always there.

Around the same time Wilson and Penzias discovered this mysterious radiation, a group of theoretical astrophysicists at Princeton University was working on a model of the universe in which the universe expands and contracts in turn. This group theorized that the universe is currently expanding from an extremely hot, dense state, which would have produced radiation that might still be measurable.

In 1965 Wilson and Penzias joined the Princeton group and found that their measurements of the wavelength of the background radiation corresponded with the theorists' predictions. This result supports the big bang theory, which describes an explosion that gave rise to all the matter and radiation in the universe.

Pyotr Kapitza

*"Soviet physicist Pyotr Leonidovich Kapitza won the 1978 Nobel Prize in physics. He is known for his work in the liquefaction of gases, especially helium and hydrogen. He also studied the effects of low temperatures and strong magnetic fields on metals."*

# Pyotr Leonidovich Kapitza

**Kapitza, Pyotr Leonidovich** (1894-1984), Russian physicist and Nobel laureate in physics. Kapitza was awarded the 1978 Nobel Prize in physics for his research in low-temperature physics. He shared the prize with American physicists Arno Penzias and Robert W. Wilson, who were recognized for their work in radio astronomy.

Kapitza was born in Kronshtadt, Russia, and educated at the Petrograd Polytechnic Institute and the Petrograd Physical and Technical Institute in what is now Saint Petersburg. For two years after his graduation in 1919 he taught electrical engineering at the Petrograd Polytechnic Institute.

In 1921 Kapitza went to England to study at the University of Cambridge as part of the renewal of scientific relations between the new Union of Soviet Socialist Republics (USSR) and the West. Soon after he arrived, Kapitza became an assistant to Sir Ernest Rutherford, the director of magnetic research at the Cavendish Laboratory in Cambridge. Kapitza earned his Ph.D. degree in physics from Cambridge in 1923.

Kapitza stayed at Cambridge for more than a decade after earning his Ph.D. degree. He worked on producing strong magnetic fields, but soon began studying the effects that these strong fields had on metals. He found that the magnetic properties of metals in high magnetic fields grew more interesting at very low temperatures.

Kapitza returned to the USSR in 1934 and was refused permission to leave. In 1936 he became director of the Institute for Physical Problems of the USSR Academy of Sciences, and Soviet leader Joseph Stalin had Kapitza's Cambridge laboratory moved to Moscow. In Moscow in 1941 Kapitza first published his findings on the superfluidity of helium II. When helium is cooled to about -271° C (-455° F), it becomes a better conductor than copper and it flows even more easily than gases. It can climb the walls of a container and seep through a sealed lid. Kapitza's work on helium II earned him the 1978 Nobel Prize.

Kapitza refused to work on the Soviet atomic weapons program and was placed under house arrest from 1945 to 1953. After Stalin's death, Kapitza resumed his place at the Institute for Physical Problems. There he studied subjects ranging from ball lightning to solid state physics.

Steven Weinberg

*"American physicist Steven Weinberg won the 1979 Nobel Prize in physics. He was awarded the prize for his study of the interactions of atomic particles."*

## STEVEN WEINBERG

**Weinberg, Steven** (1933-), American physicist and Nobel laureate. Born in New York City, Weinberg graduated from cornell University in 1954, attended Copenhagen's Nordic Institute for Theoretical Atomic Physics for a year, and obtained his doctorate from Princeton University in 1957. He taught at the University of California at Berkeley, the Massachusetts Institute of Technology and, from 1973, at Harvard University. In 1967, together with the Pakistani physicist Abdus Salam, Weinberg offered a hypothesis that unified the known facts about the electromagnetic and the weak interactions between atomic particles. When this so-called unification hypothesis was later tested experimentally, the outcomes it predicted proved true, unlike those of a number of alternative hypotheses. In 1979 the two physicists shared the Nobel Prize in physics with the American physicist Sheldon Lee Glashow for their contribution to the understanding of the interactions of elementary particles.

Sheldon Lee Glashow

*"American physicist Sheldon Lee Glashow won the 1979 Nobel Prize in physics. He developed a theory that identified electromagnetic interactions between elementary particles caused by weak nuclear force."*

# SHELDON LEE GLASHOW

**Glashow, Sheldon Lee** (1932-), American physicist and Nobel laureate, who helped to span what appeared to be divergent explanations of the physical world and advance physics toward a single unifying theory. Born in New York City, Glashow went to Bronx High School of Science, Cornell University, and Harvard University, where he received a Ph.D. in physics in 1958. After five years at the University of California in Berkeley, he returned to Harvard to work on a theory involving the unification of weak interactions and electromagnetism. In 1979 he shared the Nobel Prize in physics with the physicists Steven Weinberg of Harvard and Abdus Salam of the University of London for devising a theory that demonstrates the identity of electromagnetic interactions of particles and interactions caused by weak nuclear force.

Abdus Salam

*"Pakistani physicist Abdus Salam won the 1979 Nobel Prize in physics. He won the award for his work in developing a unification hypothesis concerning electromagnetic and weak interactions between atomic particles."*

# ABDUS SALAM

**Salam, Abdus** (1926-1996), Pakistani physicist and Nobel laureate, known for his contributions to the understanding of the interactions of elementary particles. Salam was born in Jhang Sadar, India (now in Pakistan), attended the Government College at Lahore, and received a doctorate in mathematics and physics from the University of Cambridge in 1952. He taught at both institutions before becoming professor of theoretical physics at Imperial College, London, in 1957, and he was made director of the International Centre for Theoretical Physics in Trieste, Italy, when it was established in 1964. In 1967, with the American physicist Steven Weinberg, Salam offered a so-called unification hypothesis that incorporated the known facts about the electromagnetic and weak interactions between atomic particles Elementary Particles. When tested, the hypothesis held up, unlike a number of alternative hypotheses. The men shared the 1979 Nobel Prize in physics for this work with American physicist Sheldon Lee Glashow, who also contributed to the understanding of particle interactions.

James Cronin

*"American physicist James Cronin won the 1980 Nobel Prize in physics for demonstrating the changes that occur when matter transforms into antimatter."*

# James Watson Cronin

**Cronin, James Watson** (1931- ), American physicist and Nobel laureate. Cronin shared the 1980 Nobel Prize for physics with American Val Logsdon Fitch for demonstrating that, unlike what was previously thought, symmetry is not always preserved when some elementary particles change in state from matter to antimatter.

Cronin was born in Chicago, Illinois, and received his bachelor's degree at Southern Methodist University in 1951. He earned both his master's and doctoral degrees in 1955 in physics from the University of Chicago. From 1955 to 1958 Cronin served as an assistant physicist at Brookhaven National Laboratory in Long Island, New York. He taught at Princeton University from 1958 to 1971 and in 1971, Cronin became professor of physics at the University of Chicago.

In 1964, Cronin, Fitch, and some of their colleagues were studying a subatomic particle called the K. meson, also known as a kaon. A meson is a particle that weighs approximately half as much as a proton. The kaon is an unstable, neutral (uncharged) meson that had been discovered in the interactions between particles from outer space and particles from the earth's atmosphere.

Physicists thought for some time that the universe followed three fundamental rules of symmetry. The first, the symmetry of charge conjugation (C), states that the result of an experiment should not change if all the particles in the experiment are switched from antimatter to matter or vice versa. The second rule of symmetry, parity (P), says that the result of an experiment should not change if the positions of all the particles in the experiment are completely reversed. The third rule, time-reversal (T), says that any reaction between elementary particles should occur equally well in either direction; that is, if two particles can come together to form one particle, the resulting particle should be able to decay to form the original two particles.

Before Cronin and Fitch started their experiment, the rules C and P had both been found to be untrue in some cases. Physicists tried to preserve the idea of symmetry by theorizing that any reaction had to be symmetric over a combination of CP; that is, if C symmetry was violated, P symmetry would have to be violated, too. Cronin and Fitch disproved this rule by showing that kaons do not preserve P, C, or CP symmetry some of the time. However, one rule of symmetry remains above suspicion: all reactions must be symmetric in the combination of charge conjugation, parity, and time (CPT).

Since every reaction has to be symmetric under CPT, Cronin and Fitch's result showed that T symmetry has to fail when CP symmetry fails. That means reactions that do not preserve CP symmetry cannot run backward. Some scientists think this explains how more matter than antimatter is created in the universe.

Val Fitch

*"American theoretical physicist Val Fitch won the Nobel Prize in physics in 1980. Fitch showed that the K.-mesons resulting from proton collisions did not obey the absolute principle of symmetry".*

# VAL LOGSDON FITCH

**Fitch, Val Logsdon** (1923- ), American experimental physicist and Nobel Prize winner. Fitch is noted for his studies of subatomic particles and for the discovery that one of the elementary particles, the neutral K-meson (now called the kaon), does not always follow the principles of symmetry, which were once thought to be universal traits. For their discovery of the violations of fundamental symmetry principles in the decay of neutral K-meson particles, Fitch and American physicist James Watson Cronin together were awarded the 1980 Nobel Prize in physics.

Fitch was born on a cattle ranch in Cherry County, Nebraska. His career as a physicist began during World War II (1939-1945) when, as a United States soldier, he was stationed in Los Alamos, New Mexico, to work on the Manhattan Project to design and build the atomic bomb. He developed excellent research skills and worked with future Nobel Prize winners, although he still had to complete his undergraduate degree, which he did in 1948 at McGill University in Montréal, Québec, Canada. Fitch earned his Ph.D. degree in 1954 at Columbia University in New York City. That same year, he joined the faculty at Princeton University in New Jersey, where he remained throughout his career.

At Princeton, Fitch teamed with Cronin to explore the characteristics and behavior of subatomic particles. Until the 1950s physicists believed that there was perfect balance, or symmetry, between matter and antimatter. This symmetry was described as CP, both a balance of positive and negative electrical charges (C) and parity, or equality, between left-handed and right-handed orientation (P). Fitch and Cronin found that this is not always true. In 1964 they observed that on rare occasions, the decay of neutral K-meson particles violates CP symmetry. The scientists modified the CP symmetry principle and confirmed CPT (charge, parity, time-reversal) symmetry, where there is a balance between matter and antimatter moving forward and backward in time, respectively.

Fitch and Cronin's research forced physicists to reexamine a number of theories. In particular, their discovery may explain the formation of sufficient matter to create our universe following the theoretical explosion known as the big bang. Without the violation of CP symmetry, matter and antimatter would have canceled each other out, and the big bang would have produced only gamma radiation, not our known universe.

**Nicolaas Bloembergen**

*"Dutch-American physicist Nicolaas Bloembergen won the Nobel Prize in physics in 1981. Bloembergen developed laser spectroscopy."*

## Nicolaas Bloembergen

**Bloembergen, Nicolaas** (1920- ), Dutch-American physicist and Nobel Prize winner. Bloembergen is noted for his pioneering research in laser spectroscopy, a technique that uses energy emissions to study the properties of matter. For his work in developing laser spectroscopy, Bloembergen received the 1981 Nobel Prize in physics, which he shared with Swedish physicist Kai Manne Borje Siegbahn and American physicist Arthur Leonard Schawlow.

Bloembergen was born in Dordrecht, the Netherlands, and received his Ph.D. degree in 1948 from the Leiden University. In 1951 he joined the faculty of Harvard University in Cambridge, Massachusetts, where he spent the remainder of his career. He became a United States citizen in 1958.

Spectroscopy is the study of the electromagnetic spectrum produced by a substance when exposed to certain kinds of energy, such as radiation. The substance absorbs or emits some of the energy, thereby producing a spectrum that can be carefully measured and analyzed. The spectrum provides information about molecular-energy levels, chemical bonds, and other features of the substance.

Bloembergen was especially interested in using lasers to excite a substance, and then studying the relative amounts of energy the substance absorbs. Lasers are intense beams of light waves. However, at very high intensities, the traditional laws of optics do not apply. Bloembergen worked out new laws of optics for these situations and used these laws to develop additional techniques for laser spectroscopy. Applications for these techniques range from the analysis of biological substances to the study of combustion in jet engines.

Arthur Schawlow

*"American physicist Arthur Schawlow won the 1981 Nobel Prize in physics. He was awarded the prize for his work in developing laser spectroscopy."*

# Arthur Leonard Schawlow

**Schawlow, Arthur Leonard** (1921-1999), American physicist and Nobel Prize winner. His research focused on optics, in particular, lasers and their use in spectroscopy. For his work in the development of laser spectroscopy, Schawlow was honored with the 1981 Nobel Prize in physics, which he shared with Dutch-American physicist Nicolaas Bloembergen and Swedish physicist Kai Manne Borje Siegbahn.

Born in Mount Vernon, New York, Schawlow was educated in Canada, receiving his Ph.D. degree in 1949 from the University of Toronto. He worked for several years at Columbia University in New York City and then at Bell Telephone Laboratories. In 1961 he joined the faculty of Stanford University in Palo Alto, California, where he spent the remainder of his career.

A noted author and educator, Schawlow made contributions in a number of areas of physics, including superconductivity, nuclear resonance, and spectroscopy. He was especially instrumental in developing lasers, which are intense beams of light waves all of the same wavelength. Working first with microwaves, which have longer wavelengths than visible light, Schawlow described microwave spectroscopy in a classic text he coauthored with Charles Townes (who shared the 1964 Nobel Prize in physics).

In the 1950s Schawlow described organized wavelengths in the optical region-that is, lasers. This contributed to the first successful generation of a laser, achieved in 1960 by American physicist Theodore Maiman. Schawlow saw the potential usefulness of lasers in spectroscopy, which is the study of the electromagnetic spectrum that a substance produces when exposed to certain kinds of energy, such as radiation. The substance absorbs or emits some of the energy, thereby producing a spectrum that can be carefully measured and analyzed. The spectrum provides information about molecular-energy levels, chemical bonds, and other fundamental features of the substance.

Kai Siegbahn

*"Swedish physicist Kai Siegbahn won the 1981 Nobel Prize in physics. He was awarded the prize for his work in developing high-resolution electron spectroscopy."*

# Kai Manne Borje Siegbahn

**Siegbahn, Kai Manne Borje** (1918- ), Swedish physicist and Nobel Prize winner. He was instrumental in developing high-resolution spectroscopy, a technique used to study characteristics of matter, such as atomic structure, chemical bonds, and other features. Siegbahn developed high-resolution electron spectroscopy, for which he was awarded the 1981 Nobel Prize in physics, shared with Dutch-American physicist Nicolaas Bloembergen and American physicist Arthur Leonard Schawlow, who were honored for their work in laser spectroscopy.

Born in Lund, Siegbahn is the son of Karl Manne Georg Siegbahn, who received the 1924 Nobel Prize in physics. Kai Siegbahn attended the University of Stockholm, where he received his Ph.D. degree in 1944. He taught at the Royal Institute of Technology and has continued to teach at the University of Uppsala.

Spectroscopy is the study of the electromagnetic radiation that a substance produces when exposed to certain kinds of energy, such as radiation. The substance absorbs or emits some of the energy, thereby producing a spectrum that can be carefully measured and analyzed. Before the 1950s electron spectroscopy had various limitations. When a substance was excited with radiation, the energy of the electrons it emitted could not be analyzed in its entirety. Siegbahn in 1954 developed a type of spectrometer that provided a more accurate analysis of the electrons' energy and, consequently, a more complete electromagnetic spectrum. This tool enabled Siegbahn and his coworkers to study characteristics of electrons in almost all the known chemical elements and also contributed to the development of new, high-tech materials, such as polymers, plastics, and medical devices.

Kenneth G. Wilson

*"American physicist Kenneth G. Wilson won the 1982 Nobel Prize in physics. He developed a theory describing how matter undergoes phase transition when critical phenomena occur such as ice melting from solid to liquid at a critical temperature."*

# Kenneth Geddes Wilson

**Wilson, Kenneth Geddes** (1936- ), American physicist and Nobel Prize winner. Wilson is best known for his contribution to the understanding of how bulk matter undergoes phase transition (sudden and profound structural changes resulting from variations in environmental conditions; see Matter, States of). His theory, for which he received the 1982 Nobel Prize in physics, offers an explanation for diverse phenomena including ice melting and iron losing its magnetism at certain critical temperature or pressure levels.

Wilson was born in Waltham, Massachusetts. His scientific prowess was recognized early-he entered Harvard University at the age of 16, receiving his B.S. degree in math and physics in 1956. He earned his Ph.D. degree in theoretical physics at the California Institute of Technology (Caltech) in 1961 under the tutelage of Nobel Prize winner Murray Gell-Mann. From 1959 to 1962 Wilson served as a Harvard junior fellow, followed by a year as a Ford Foundation fellow working in the field of elementary particles at the European Center for Nuclear Energy Research (CERN, now the European Laboratory for Particle Physics) in Geneva, Switzerland.

In 1963 Wilson joined Cornell University's Department of Physics in Ithaca, New York, where he first became interested in the study of phase transitions, then a new branch of theoretical physics. He was promoted from assistant to full professor in 1970, and was selected to fill the James A. Weeks Chair in Physical Sciences at Cornell in 1974. In 1985 he became the director of Cornell's Center for Theory and Simulation in Science and

Engineering. In 1988 Wilson moved to Ohio State University to become the Hazel C. Youngberg Trustees Distinguished Professor, a post he holds today.

In the phase-transition theory scientists had been exploring why substances change their physical state when subjected to certain temperatures or pressure levels. A simple illustration of phase transition is the melting of ice-a transition of water from solid to liquid that occurs at a critical temperature. Another example is the behavior of an iron magnet-if the iron is heated to a certain degree, it suddenly loses its magnetic properties. Wilson sought to identify the commonality in these seemingly diverse "critical phenomena." Applying a mathematical technique called *renormalization*, Wilson developed a description of substance behavior when it is close to the critical point, along with a method to calculate numerically the quantities crucial for phase transition. His phase-transition theory allows scientists to accurately predict critical phenomena, and has application in fields ranging from chemistry to engineering.

Subrahmanyan Chandrasekhar

*"Indian-born American theoretical astrophysicist Subrahmanyan Chandrasekhar won the 1983 Nobel Prize in physics. He studied the process by which stars evolve."*

## SUBRAHMANYAN CHANDRASEKHAR

**Chandrasekhar, Subrahmanyan** (1910-1995), American theoretical astrophysicist and Nobel laureate, who contributed greatly to the current understanding of stellar evolution. He was born in Lahore, India (now Pakistan), and was educated in India and at Trinity College, University of Cambridge, earning a Ph.D. in 1933. In 1953 he became a U.S. citizen.

Although Chandrasekhar worked on theories of radiative transfer and convective transport of heat in stellar atmospheres, his most important studies concerned the small, dim, hot, dense stars known as white dwarfs (see Star). He determined that a star with a mass more than 1.44 times that of the mass of the sun cannot directly become a white dwarf, a limit now called the Chandrasekhar limit. He shared the Nobel Prize for physics in 1983 with U.S. physicist William A. Fowler for his work on stars. His books include An Introduction to the Study of Stellar Structure (1939) and Principles of Steller Dynamics (1942). The Chandra X-ray Observatory, a powerful telescope that the United States plans to launch into Earth's orbit as a satellite in 1999, was named after Chandrasekhar.

William Fowler

*"American physicist William Fowler won the 1983 Nobel Prize in physics. He studied the nuclear synthesis of the chemical elements within stars."*

## William Fowler

**Fowler, William** (1911-1995), American physicist and astronomer, born in Pittsburgh, Pennsylvania, who made major contributions to cosmology theory. Fowler obtained his doctorate in 1936 from the California Institute of Technology, where he later taught and conducted research on stellar processes. The work for which he is best known was developed in collaboration with the British astronomers Fred Hoyle and Geoffrey and Margaret Burbidge and published in 1957. It described the processes of nuclear synthesis of chemical elements within stars, and it has become a cornerstone of modern astrophysics. Fowler shared the 1983 Nobel Prize for physics with the astrophysicist Subrahmanyan Chandrasekhar.

Carlo Rubbia

*"Italian physicist Carlo Rubbia won the 1984 Nobel Prize in physics. He won the award for his discovery of subatomic W and Z particles."*

## CARLO RUBBIA

**Rubbia, Carlo** (1934-), Italian physicist, who shared the 1984 Nobel Prize in physics with the Dutch physicist Simon van der Meer for their discovery of the subatomic W and Z particles. The W and Z particles transmit the weak force, one of the four fundamental forces of nature (the other three are gravity, electromagnetism, and the strong force). Rubbia was born in Gorizia, and obtained his doctorate from the University of Pisa in 1957. In 1960 he became a physicist at the European Center for Nuclear Energy Research (CERN, now the European Laboratory for Particle Physics) in Geneva. In 1970 he also was appointed a physics professor at Harvard University, and he continued in both positions from that time. The confirmed observation of the W and Z particles in 1983 by Rubbia and his colleagues was predicted by and further substantiated, the electroweak theory, which unites the electromagnetic and the weak nuclear forces.

Simon van der Meer

*"Dutch physicist Simon van der Meer won the 1984 Nobel Prize in physics. His work contributed significantly to the discovery of subatomic particles W and Z, supporting Einstein's theory that all forces in nature are related."*

# SIMON VAN DER MEER

**van der Meer, Simon** (1925- ), Dutch physicist and co-winner of the 1984 Nobel Prize for physics for his contributions to the discovery of several subatomic particles whose existence had been predicted but not confirmed. This discovery added further evidence to the theory proposed by Albert Einstein that all forces in nature are related and helped explain the nuclear reactions that occur in the sun. Van der Meer shared the Nobel Prize with his colleague Italian physicist Carlo Rubbia.

Born in The Hague, the Netherlands, Van der Meer obtained an engineering degree in 1952 from the Technische Hogeschool in Delft. After working for several years in the electronics industry, he joined the European Center for Nuclear Research (CERN) in Switzerland, where he spent most of his career.

Since the 1950s, theoretical physicists have searched for evidence to support Einstein's unification theory, in which all of nature's forces (such as magnetism and gravity) are related. One piece of necessary evidence is the existence of W and Z field particles. W and Z particles are about 100 times heavier than the proton, which is found in the atomic nucleus. The existence of these particles had been predicted but never confirmed. In an experiment designed by Rubbia, researchers hoped to observe W particles, with a positive or negative charge, and Z particles by colliding a beam of protons with a beam of antiprotons (protons with a negative charge). These particles convey the "weak force," a force that causes certain particles to decay, or transform into other particles. Van der Meer's most significant contribution to this effort was developing a way to create the concentrated beam of protons and antiprotons essential for the experiment. Van der Meer built a device that could generate and store antiprotons, an especially difficult task. In January of 1983 a team of more than 100 physicists confirmed for the first time the existence of W and Z particles.

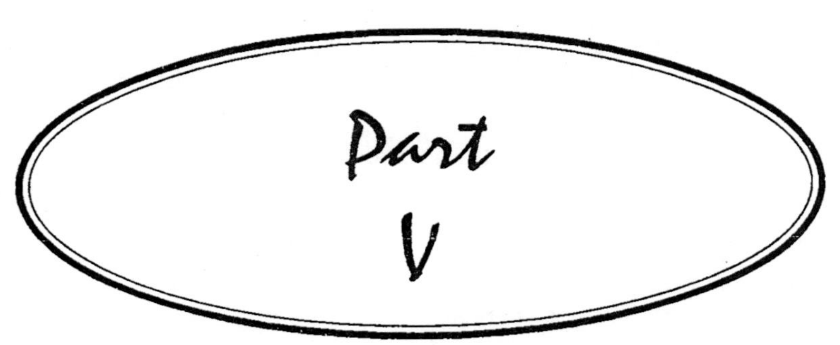
Part
V

*1985* *Klaus von Klitzing (West Germany)*
*1986* *Ernest Ruska (West Germany)*
*Gerd Binnig (West Germany)*
*Heinrich Rohrer (Switzerland)*
*1987* *K. Alex Muller (Switzerland)*
*J. Georg Bednorz (Switzerland)*
*1988* *Leon Max Lederman (United States)*
*Melvin Schwartz (United States)*
*Jack Steinberger (United States)*
*1989* *Hans G. Dehmelt (United States)*
*Wolfgang Paul (West Germany)*
*Norman F. Ramsey (United States)*
*1990* *Richard E. Taylor (Canada)*
*Jerome I. Friedman (United States)*
*Henry W. Kendall (United States)*
*1991* *Pierre Gilles de Gennes (France)*
*1992* *Georges Charpak (France)*
*1993* *Joseph H. Taylor (United States)*
*Russell A. Hulse (United States)*
*1994* *Clifford G. Shull (United States)*
*Bertram N. Brockhouse (Canada)*
*1995* *Martin L. Perl (United States)*
*Frederick Reines (United States)*
*1996* *David M. Lee (United States)*
*Douglas D. Osheroff (United States)*
*Robert C. Richardson (United States)*
*1997* *Steven Chu (United States)*
*Claude Cohen-Tannoudji (France)*
*William D. Phillips (United States)*
*1998* *Robert Laughlin (United States)*
*Daniel Tsui (United States)*
*Horst Störmer (Germany)*
*1999* *Gerardus 't Hooft (Netherlands)*
*Martinus J.G. Veltman (Netherlands)*
*2000* *Zhores I. Alferov (Russia)*
*Herbert Kroemer (Federal Republic of Germany)*
*Jack S. Kilby (United States)*
*2001* *Eric A. Cornell (United States)*
*Wolfgang Ketterle (Federal Republic of Germany)*
*Carl E. Wieman (United States)*
*2002* *Raymond Davis (United States)*
*Masatoshi Koshiba (Japan)*
*Riccardo Giacconi (United States)*

Klaus von Klitzing

*"German physicist Klaus von Klitzing won the 1985 Nobel Prize in physics. He discovered that electrical resistance occurs in precise units, rather than by a continuous and smooth process."*

# Klaus-Olaf von Klitzing

**von Klitzing, Klaus-Olaf** (1943- ), German physicist and Nobel Prize winner. Von Klitzing made a significant contribution to the field of semiconductor electronics by discovering the quantized Hall effect, an exact way of measuring electrical conductivity. His finding, for which he received the 1985 Nobel Prize in physics, allows more accurate testing of theories about electronic movements within atoms and makes possible the establishment of a new international standard for the ohm, the unit of measure of electrical resistance.

Born in Schroda, Germany (now part of Poland), von Klitzing fled westward to Lutten, Germany, with his family and millions of his countrymen to escape the invasion of the Soviet Army into Germany at the end of World War II (1939-1945). He earned his B.S. degree in physics at the Technical University in Brunswick in 1969, and his Ph.D. degree at the University of Wurzbürg in 1972. From 1975 to 1976 he was affiliated with the University of Oxford in England, where he conducted research utilizing the powerful superconducting magnets manufactured there. In 1978 he returned to the University of Würzburg before moving to the High-Field Magnet Laboratory of the Intstitut Max von Laue-Paul Langevin in Grenoble, France, where he made the discovery that earned him the Nobel Prize. For more than a decade, von Klitzing has been the director of the Max Planck Institute for Solid State Research in Stuttgart, Germany.

The Hall effect, discovered in the 19th century by American physicist Edwin Hall, is a phenomenon in which an electrical current is passed through a conducting material while a magnetic field is applied at right angles to the current. This results in an accumulation of electrons, called the *Hall voltage*, along one edge of the conductor. Von Klitzing discovered that by using extremely powerful magnetic fields and temperatures close to absolute zero (-273° C, -460° F), the Hall voltage varies in a step pattern as the magnetic field or the electric current is varied. The electrical resistance occurs in extrememly precise units regardless of the material through which the electric current is flowing-in other words, any type or thickness of wire can be used to the same effect.

Von Klitzing's discovery permits more precise measurement of electrical resistance and more accurate testing of theories about electrical movements within atoms. It also opens the door to a new and more exact standard for the ohm, the unit of electrical resistance, as well as the manufacture of equipment capable of higher electronic precision.

Ernst Ruska

*"German electrical engineer Ernst Ruska won the 1986 Nobel Prize in physics. His efforts in building the first electron microscope and other contributions in electron optics won him the award."*

# Ernst August Friedrich Ruska

**Ruska, Ernst August Friedrich** (1906-1988), German physicist, electrical engineer, and Nobel Prize winner. For his design of the first electron microscope, and subsequent improvements to the instrument, Ruska shared the 1986 Nobel Prize in physics with physicists Gerd Karl Binnig of Germany and Heinrich Rohrer of Switzerland, who designed the scanning tunneling microscope (STM).

Born in Heidelberg, Ruska earned an engineering degree from the University of Berlin in 1931 and a Ph.D. degree in 1933. In 1937 Ruska accepted a research position at Siemens-Reiniger-Werke. He remained there until 1955, when he became director of the Institute for Electron Microscopy of the Fritz Haber Institute. Concurrently, Ruska served at the institute and as professor at the Technical University of Berlin, from 1957 until his retirement in 1972.

Before leaving the University of Berlin in the 1930s, Ruska had already made the contribution to science that would bring him fame. He knew that electrons possess a wave aspect, so he believed he could treat them in a fashion similar to light waves. Ruska was also aware that magnetic fields could manipulate electrons, possibly focusing them as optical lenses do light. After confirming these principles through research, he set out to design an electron microscope. Ruska had deduced that an electron microscope would be much more powerful than an ordinary optical microscope, because he knew that magnification increased with shorter reflective waves. Since electron waves were shorter than ordinary light waves, it followed that they would allow for greater magnification. In 1932 Ruska and a collaborator, German physicist Max Knoll, under whom he obtained his doctorate, built the first crude electron microscope. Despite the fact that it was primitive and not fit for practical use, the instrument was still capable of magnifying objects 400 times.

Although modern electron microscopes can magnify an object 2 million times, they are still based upon Ruska's prototype and his correlation between wavelength and magnification. The electron microscope is an integral part of many laboratories. Researchers use it to examine biological materials (such as microorganisms and cells), a variety of large molecules, medical biopsy samples, metals and crystalline structures, and the characteristics of various surfaces.

Gerd Binnig

*"German physicist Gerd Binnig won the Nobel Prize in physics in 1986. Binnig worked on the design of the scanning tunnelling microscope with swiss physicist Heinrich Rohrer."*

# GERD KARL BINNIG

**Binnig, Gerd Karl** (1947- ), German physicist and Nobel Prize winner. He invented, with colleague Swiss physicist Heinrich Rohrer, the scanning tunneling microscope (STM), a new type of powerful microscope capable of detecting images at the atomic level. For this accomplishment, they shared the 1986 Nobel Prize in physics with German physicist Ernst August Friedrich Ruska, who was honored for his invention of the electron microscope.

Born in Frankfurt, Binnig was educated in that city at the J. W. Goethe University, where he earned his Ph.D. degree in 1978. He joined the International Business Machines (IBM) Research Laboratory near Zürich, Switzerland, that same year, and began working with Rohrer on a problem that required information on a microscopically small surface. They developed the idea of a probe that could move across an object's surface to obtain this information. The ultimate result was the creation of the scanning tunneling microscope.

The STM that Binnig and Rohrer invented is based on the wavelike property of electrons, first identified in the 1920s by Nobel laureate Louis Victor de Broglie. The microscope's sharp probe moves near the sample's surface, and any change in distance between the probe and the sample-even as small as the diameter of a single atom-is recorded by the microscope. By moving the probe in a sweeping motion, a three-dimensional image can be produced. This image shows detail not possible with any other kind of microscope. It can reveal the surface of a material at the atomic level and can also provide information about atomic composition. The STM has been used to study biological samples, to analyze industrial materials (such as superconductors, materials that conduct electric current with no resistance at temperatures near zero), and to test miniaturized electronic circuits.

Heinrich Rohrer

*"Swiss physicist Heinrich Rohrer won the 1986 Nobel Prize in physics. Rohrer shared credit for inventing the scanning tunnelling microscope, a powerful microscope capable of producing three dimensional images of materials at the atomic level."*

## HEINRICH ROHRER

**Rohrer, Heinrich** (1933- ), Swiss physicist and co-winner of the 1986 Nobel Prize for physics for his invention of the scanning tunneling microscope, a new type of powerful microscope capable of detecting images at the atomic level. Rohrer shared the prize with German physicists Gerd Karl Binnig and Ernst August Friedrich Ruska.

Rohrer was born in Buchs, Sankt Gallen, Switzerland, and studied at the Swiss Federal Institute of Technology, where he earned his Ph.D. degree in 1960. He joined the IBM Research Laboratory near Zürich, Switzerland, in 1963. There, he and Binnig turned their attention to an experiment that required the study of a microscopic surface. In so doing they created a new type of microscope.

The scanning tunneling microscope is based on the wavelike property of electrons. The microscope has a sharp probe that moves near the sample's surface in a vacuum, and emits electrons. The electrons tunnel, or flow through the vacuum, from the probe's tip to the sample's surface. The microscope records any change in distance between the probe and the sample-even as small as the diameter of a single atom. By moving the probe in a sweeping motion, a three-dimensional image of the sample can be produced. This image shows detail not possible with any other kind of microscope. It can reveal the surface of a material at the atomic level and also provide information about its atomic composition. The scanning tunneling microscope has been used to study biological samples, to analyze industrial materials (such as superconductors), and to test miniaturized electronic circuits.

Karl Muller

*"Swiss physicist Karl Müller won the 1987 Nobel Prize in physics. Müller discovered that certain materials can become superconductive (able to conduct an electrical current without resistance) at much higher temperature than once thought possible."*

# KARL ALEXANDER MÜLLER

**Müller, Karl Alexander** (1927- ), Swiss physicist and cowinner of the 1987 Nobel Prize in physics for his discovery that copper oxide ceramic materials can achieve *superconductivity* (the ability of a material to carry an electrical current indefinitely without resistance) at temperatures well above the extremely low temperatures once associated with this remarkable property. Müller shared the prize with his colleague, German physicist Georg Johannes Bednorz.

Müller was born in Switzerland. He attended the Swiss Federal Institute of Technology, receiving his Ph.D. degree in physics there in 1958. He worked at the Battelle Institute in Geneva for several years and then in 1963 joined the International Business Machines (IBM) Research Laboratory near Zürich, Switzerland, where he spent most of his career. Müller was also on the faculty of the University of Zürich.

Superconductivity had been discovered in 1911. However, practical applications could not be developed because to achieve superconductivity materials had to be cooled to temperatures close to absolute zero (0 K, -273° C, -459° F). In the 1970s compounds containing the element niobium were found to be superconducting at 23 K (-250° C, -418° F), which theoretical physicists believed to be the upper temperature limit for superconductors. By systematically testing oxides containing nickel or copper, in 1986 Bednorz and Müller found a material, barium lanthanum copper oxide, that starts to become superconducting at temperatures higher than 100 K (-173° C, -279° F). This temperature is easier and cheaper to maintain, making it possible for researchers to design superconducting devices. Superconductors are now used in scientific and medical instruments and may find applications in the electronics industry and in electric power transmission and storage.

J. Georg Bednorz

*"German physicist Johannes Bednorz won the Nobel Prize in physics in 1987. His work involved the discovery of compounds that can achieve superconductivity at high temperatures."*

# Johannes Georg Bednorz

**Bednorz, Johannes Georg** (1950- ), German physicist and Nobel Prize winner. Bednorz and his colleague, Swiss physicist Karl Alex Müller, discovered that copper oxide ceramic materials can achieve superconductivity (the ability of a material to carry an electrical current indefinitely without resistance) at temperatures well above the extremely low temperatures once associated with this remarkable property. For this discovery, Bednorz and Müller shared the 1987 Nobel Prize in physics.

Bednorz was educated at the University of Münster and the Swiss Federal Institute of Technology, where he received his Ph.D. degree in 1982 under the guidance of Müller. That same year, Bednorz joined the staff of the International Business Machines (IBM) Research Laboratory near Zürich, Switzerland.

Shortly after he completed his doctoral work, Bednorz joined Müller in an investigation of superconducting materials that were discovered in 1911 by Dutch physicist Heike Kamerlingh Onnes, winner of the 1913 Nobel Prize in physics. These materials had to be cooled to temperatures close to absolute zero (0 K, -273° C, -459° F) to achieve superconductivity, making them impractical for industrial purposes. In the 1970s compounds containing the element niobium were found to be superconducting at 23 K (-250° C, -418° F), which physicists believed to be the upper temperature limit for superconductors. This temperature was still too cold for practical applications. By systematically testing nickel- and copper-containing oxides, Bednorz and Müller found in 1986 a material, barium lanthanum copper oxide, that starts to become superconducting at temperatures higher than 100 K (-173° C, -279° F). Now a practical reality, superconductors are used in scientific and medical instruments, and may find applications in the electronics industry and in electric-power transmission and storage.

Leon Lederman

*"American physicist Leon Lederman won the 1988 Nobel Prize in physics. He helped discover the muon neutrino, one of the elementary particles of matter."*

# LEON MAX LEDERMAN

**Lederman, Leon Max** (1922- ), American physicist and cowinner of the 1988 Nobel Prize for physics for his discovery of the muon neutrino, one of the elementary particles that make up the atom, proving the existence of more than one type of neutrino. Lederman shared the prize with his colleagues American physicists Melvin Schwartz and Jack Steinberger.

Lederman attended the City College of New York and then Columbia University, where he earned his Ph.D. degree in 1951. He joined the faculty at Columbia and directed its Nevis Laboratory from 1961 to 1978. In 1979 he became the director of the Fermi National Accelerator Laboratory in Batavia, Illinois, near Chicago and served in that position for ten years. He joined the faculty of the University of Chicago in 1989 and later the Illinois Institute of Technology in Chicago.

Neutrinos and other elementary particles, the basic building blocks of matter, are smaller than the particles found in the atomic nucleus (protons and neutrons). As a result, they are difficult to isolate and study. In the early 1960s Lederman, Schwartz, and Steinberger devised a way to capture neutrinos. Using the powerful particle accelerator at the Brookhaven National Laboratory in New York, they created a beam of high-energy neutrinos. With a specialized detector, Lederman and his colleagues were able to study the neutrinos and, in doing so, found that neutrinos exist in more than one variety. Their discovery of the muon neutrino inspired other physicists to search-often successfully-for additional particles.

In addition to his Nobel Prize-winning research, Lederman discovered in 1956 the long-lived neutral K-meson particle. In 1977 he found evidence for yet another elementary particle, the bottom quark.

Melvin Schwartz

*"American physicist Melvin Schwartz won the 1988 Nobel Prize in physics. He was awarded the prize for discovering the muon neutrino, an elementary particle."*

# Melvin Schwartz

**Schwartz, Melvin** (1932- ), American physicist and co-winner of the 1988 Nobel Prize for his discovery of the muon neutrino (one of the elementary particles that make up the atom), proving the existence of more than one type of neutrino. Schwartz shared the prize with his colleagues American physicists Leon Max Lederman and Jack Steinberger.

Schwartz was educated at Columbia University, New York, where he studied physics as an undergraduate and graduate student. He received his doctoral degree in 1958 under Steinberger, who later became a research colleague. Schwartz spent ten years on the faculty at Columbia and then accepted a position at Stanford University in Palo Alto, California, in 1968. In 1979 he left teaching to found a business that designed security systems for computer data networks. He returned to Columbia University in 1991, also working for several years at the Brookhaven National Laboratory in New York.

Neutrinos and other elementary particles, the basic building blocks of matter, are smaller than the particles found in the atomic nucleus (protons and neutrons). As a result, they are difficult to isolate and study. In the early 1960s Steinberger, Lederman, and Schwartz devised a way to capture neutrinos and use them to discover yet other particles. Using the powerful particle accelerator at the Brookhaven National Laboratory in New York, they created a beam of charged, high-energy subatomic particles called pions that decayed into other subatomic particles called muons, thereby releasing a beam of high-energy neutrinos. With a specialized detector, the team studied the neutrinos and found that neutrinos exist in more than one variety. Their discovery of the muon neutrino inspired other physicists to search-often successfully-for additional elementary particles.

Jack Steinberger

*"American physicist Jack Steinberger won the 1988 Nobel Prize in physics. He discovered the muon neutrino, proving the existence of different types of neutrinos, elementary particles that are emitted during the decay of other particles."*

## JACK STEINBERGER

**Steinberger, Jack** (1921- ), American physicist and cowinner of the 1988 Nobel Prize for physics for his discovery of the muon neutrino (one of the elementary particles that make up the atom), proving the existence of more than one type of neutrino. Steinberger shared the prize with his colleagues American physicists Leon Max Lederman and Melvin Schwartz.

Steinberger studied chemistry as an undergraduate at the University of Chicago. As a member of the United States Army he was assigned to the MIT Radiation Laboratory, which inspired his interest in physics. He returned to the University of Chicago for graduate work, earning his Ph.D. in 1948. He then conducted research at the Institute for Advanced Study in Princeton, New Jersey, at the University of California at Berkeley, and at Columbia University. In 1968 he joined the staff of the European Center for Nuclear Research (CERN) in Switzerland, where he spent the remainder of his career.

Neutrinos and other elementary particles, the basic building blocks of matter, are smaller than the particles found in the atomic nucleus (protons and neutrons). As such, elementary particles are difficult to isolate and study. In the early 1960s, Steinberger, Lederman, and Schwartz devised a way to capture neutrinos and use them to discover yet other particles. Using the powerful particle accelerator at the Brookhaven National Laboratory in New York, they created a beam of charged, high-energy, subatomic particles called charged pions that decayed into other subatomic particles called muons, thereby releasing a beam of high-energy neutrinos. With a specialized detector, the team studied the neutrinos and found that neutrinos exist in more than one variety. Their discovery of the muon neutrino inspired other physicists to search-often successfully-for additional elementary particles.

**Hans Georg Dehmelt**

*"German-born American physicist Hans Georg Dehmelt won the Nobel Prize in physics in 1989. Dehmelt developed techniques for isolating and studying individual atoms and particles."*

# HANS GEORG DEHMELT

**Dehmelt, Hans Georg** (1922- ), German American physicist and Nobel Prize winner. Dehmelt is noted for developing techniques for isolating and studying individual atoms and particles, enabling researchers to better study fundamental atomic properties. For this work, he shared the 1989 Nobel Prize in physics with German physicist Wolfgang Paul and American physicist Norman Foster Ramsey.

Born in Görlitz, Germany, Dehmelt earned his Ph.D. degree in 1950 at the University of Göttingen. He was a research fellow for several years at Hans Kopfermann's Institute in Göttingen. In 1952 Dehmelt went to the United States, where he conducted research at Duke University. In 1955 he joined the faculty of the University of Washington, where he continues his research and teaching.

Building on the work of Wolfgang Paul, Dehmelt experimented with a three-dimensional electric field (the Paul trap) to suspend ions in a small area. By adding a strong magnetic field to the device, Dehmelt in 1973 was able to isolate and store a single electron. He used this device, called a Penning trap, to study the magnetic properties and spin states of electrons. Dehmelt also developed a technique to cool the particles being studied to slow down their movement, which improved the accuracy of the measurements being taken. These measurements proved to be so precise that they provided verification of fundamental theories in quantum theory (the science relating to mass and electromagnetic radiation in atoms).

Dehmelt continued to perfect his techniques for studying atomic particles. He improved the accuracy of his magnetic measurements in electrons, achieving an accuracy of just a few parts in a trillion. In 1980 he successfully isolated, cooled, and photographed a single ion in the Penning trap. Numerous other researchers have put this technique to use in studying the mass of atoms and atomic particles.

Wolfgang Paul

*"German physicist Wolfgang Paul won the 1989 Nobel Prize in physics. Paul becaome the first scientist to isolate individual ions, or atoms that carry an electric charge thus enabling the direct study of ions."*

# Wolfgang Paul

**Paul, Wolfgang** (1913-1993), German physicist and co-winner of the 1989 Nobel Prize for his development of a technique to isolate individual ions (atoms that have an electric charge), making it possible to study and measure them. For this work, Paul shared the 1989 Nobel Prize in physics with American physicists Hans Georg Dehmelt and Norman Foster Ramsey.

Paul was educated at the Technical Universities of Munich and Berlin, receiving his Ph.D. degree in 1939. He was on the faculty of the University of Kiel and the University of Göttingen from 1939 to 1952, when he was appointed director of the physics institute at the University of Bonn. He directed its physics institute and spent the remainder of his career, becoming professor emeritus in 1981. Paul also served as the director of the nuclear physics division of the European Center for Nuclear Research (CERN).

In the 1950s Paul experimented with electric and magnetic fields as a way to focus beams of atoms and to separate different particles. He found that a certain type of electric field, called a quadrupole electric field, could be used to separate ions with different masses. This technique became the basis of the quadrupole mass filter, an instrument eventually used in many laboratories to conduct analysis of elements by examining their individual spectrums. Paul experimented further with both electric and magnetic fields. He developed a three-dimensional version of the mass filter, which became known as the Paul trap. The Paul trap suspends ions in a small area so that researchers can study them.

Paul's other research included optical observation of the Lamb shift (a small shift in the energy levels of a hydrogen atom named after Willis Eugene Lamb Jr., its discoverer) and, with the collaboration of his two sons, the creation of a superconducting storage ring for containing and studying various slow neutrons. Paul is also credited with the development of high-energy physics in Germany through the construction of several particle accelerators in Bonn.

Norman F. Ramsey

*"American physicist Norman F. Ramsey won the 1989 Nobel Prize in physics. Ramsey dramatically improved the method of measuring atomic energy spectra, which led to the development of extremely accurate atomic clocks."*

# NORMAN FOSTER RAMSEY

**Ramsey, Norman Foster** (1915- ), American physicist and Nobel Prize winner. Ramsey, with German American physicist Hans Georg Dehmelt and German physicist Wolfgang Paul, shared the 1989 Nobel Prize in physics for inventing a highly accurate technique for studying atomic-energy oscillations, which became the basis for modern atomic clocks.

Ramsey was educated at the University of Cambridge in England, and at Columbia University in New York City, where he received his Ph.D. degree in 1940 while working with Nobel Prize-winning physicist Isidor Isaac Rabi. For several years, Ramsey worked at various institutions, including the University of Illinois, and the Massachusetts Institute of Technology (MIT) Radiation Laboratory, and then joined the faculty at Columbia University. He was a member of the group of scientists that founded Brookhaven National Laboratory in New York, and served as the first chairman of Brookhaven's physics department. In 1947 he joined the faculty of Harvard University, where he spent the remainder of his career until his retirement in 1986.

Before Ramsey's work, atomic-energy oscillations were measured by passing a beam of atoms through a constant magnetic field. However, maintaining the magnetic field at a constant level over a large area proved difficult. Ramsey used two separate magnetic fields; as a result, he achieved vastly increased accuracy in the measurements. Called the separated-oscillatory-fields method, this technique put oscillations to use in synchronizing the components of atomic clocks, providing them with their remarkable accuracy. This method also was put to use in the creation of the first atomic maser (*m*icrowave *a*mplification by *s*timulated *e*mission of radiation), which is a focused beam of microwaves (similar to the light waves that form a laser). One of the maser's many applications was in space research, where it was used to track the space probe Voyager 2.

Ramsey's other notable research includes an explanation of chemical shifts in nuclear-magnetic-resonance (NMR) experiments. These shifts in the NMR experiments were explained by the presence of electrons around the nuclei. This understanding provided the basis for magnetic-resonance-imaging (MRI) instruments used in medicine. He also studied the interactions between atomic particles, and he measured the magnetic properties of these particles.

Richard E. Taylor

*"American physicist Richard E. Taylor won the 1990 Nobel Prize in physics. Taylor and his research colleagues proved the existence of quarks the tiny particles that form the building blocks of atoms."*

# RICHARD EDWARD TAYLOR

**Taylor, Richard Edward** (1929- ), Canadian physicist and Nobel Prize winner. Taylor and his two colleagues, Henry Way Kendall and Jerome Isaac Friedman, discovered the quark, one of the fundamental building blocks of the atom. For this discovery, they shared the 1990 Nobel Prize in physics.

Taylor was educated at the University of Alberta and at Stanford University, where he received his Ph.D. degree in 1962. After a brief stay at the Lawrence Berkeley National Laboratory, he joined the faculty of the Stanford Linear Accelerator Center (SLAC) and remained there throughout his career.

Protons and neutrons, found in the atom's nucleus, were once thought to be the most basic building blocks of matter. In the 1950s physicists began to search for more-elementary particles, and in 1964 two scientists, American physicist and Nobel Prize winner Murray Gell-Mann and American physicist George Zweig independently predicted the existence of quarks. But no one had proof that these particles actually existed. Using a new and powerful particle accelerator at the SLAC, Taylor and his colleagues devised an experiment in the late 1960s-not to look for quarks, but to study protons by colliding electrons into them. The scientists expected that the electrons would pass through the protons or bounce off them, which they did at first. Then, by speeding up the electrons so that they were traveling at close to the speed of light, the researchers made a remarkable observation. Many of the electrons rebounded at odd angles, which was a completely unexpected result. Particles within the protons apparently were causing this. Additional experiments confirmed the suspicion that those particles were quarks.

The discovery of these particles inspired a great deal of additional research. Six different types of quarks have since been found, with the unusual names up, down, charm, strange, truth, and beauty. Scientists have renewed their studies on other types of particles, which are believed to be made up of quarks. They have also used this discovery to try to understand the formation of all known matter during the theoretical big bang, billions of years ago.

**Jerome Isaac Friedman**

*"American physicist Jerome Isaac Friedman won the 1990 Nobel Prize in physics. Friedman, along with his colleagues Richard Taylor and Henry Kendall, discovered the quark, a fundamental building block of matter."*

# JEROME ISAAC FRIEDMAN

**Friedman, Jerome Isaac** (1930- ), American physicist and Nobel Prize winner. Friedman and his two colleagues, Canadian physicist Richard Edward Taylor and American physicist Henry Way Kendall, discovered the quark, one of the fundamental building blocks of matter. For this discovery, they shared the 1990 Nobel Prize in physics.

Friedman received all of his higher education at the University of Chicago, earning his Ph.D. degree in 1956. After postdoctoral work there and at Stanford University, he joined the faculty at the Massachusetts Institute of Technology (MIT), where he spent the remainder of his career.

Protons and neutrons, found in the atom's nucleus, were once thought to be the elementary particles of matter. In the 1950s physicists began to search for more elementary particles, and in 1964 two scientists independently predicted the existence of quarks. But no one had proof that they actually existed. Using a new and powerful particle accelerator at the Stanford Linear Accelerator Center (SLAC), Friedman and his colleagues devised an experiment in the late 1960s to study protons by colliding electrons into them. The researchers expected the electrons to pass through the protons or to bounce off them, which they did at first. By speeding up the electrons so they were traveling close to the speed of light, the scientists made a remarkable observation. Many of the electrons rebounded at odd angles, which was a completely unexpected result. Particles within the protons apparently were causing this unusual electron behavior and additional experiments confirmed that those particles were quarks.

Six different types of quarks have since been found and have been given the names up, down, charm, strange, truth, and beauty. Scientists have renewed their studies on other types of atomic particles, which are believed to be made up of quarks. They have also used quarks to try to understand the formation of all known matter during the big bang, the theoretical gigantic explosion that led to the creation of the universe billions of years ago.

**Henry Way Kendall**

*"American physicist Herny Way Kendall won the 1990 Nobel Prize in physics. Kendall won the award for his discovery of the quark, one of the fundamental building blocks of matter."*

# HENRY WAY KENDALL

**Kendall, Henry Way** (1926-1999), American physicist and Nobel Prize winner. Kendall and his two colleagues, Canadian physicist Richard Edward Taylor and American physicist Jerome Isaac Friedman, discovered the quark, one of the fundamental building blocks of matter. For this discovery, they shared the 1990 Nobel Prize in physics.

Kendall was educated at Amherst College and the Massachusetts Institute of Technology (MIT), where he received his Ph.D. degree in 1954. After two years of postdoctoral research, he joined the faculty at Stanford University. Kendall returned to MIT in 1961, remaining on the faculty there throughout the rest of his career.

Protons and neutrons, found in the atom's nucleus, were once thought to be the elementary particles of matter. In the 1950s physicists began to search for more elementary particles, and in 1964 two scientists independently predicted the existence of a subatomic particle now called the quark. But no one had proof that they actually existed. Using a new and powerful particle accelerator at the Stanford Linear Accelerator Center (SLAC), Kendall and his colleagues devised an experiment in the late 1960s to study protons by colliding electrons into them. The researchers expected the electrons to pass through the protons or to bounce off them, which they did at first. By speeding up the electrons so they were traveling at close to the speed of light, the scientists made a remarkable observation. Many of the electrons rebounded at odd angles, which was a completely unexpected result. Particles within the protons apparently were causing this unexpected electron behavior. Additional experiments confirmed the suspicion that those particles were quarks.

Six different types of quarks have since been found and have been given the names up, down, charm, strange, truth, and beauty. Scientists have renewed their studies on other types of atomic particles, which are believed to be made up of quarks. They have also used this discovery to try to understand the formation of all known matter during the big bang, the theoretic gigantic explosion that led to the creation of the universe billions of years ago.

Pierre-Gilles de Gennes

*"Pierre-Gilles de Gennes won the Noble Prize in physics in 1991, for discovering that methods developed for studying order phenomena in simple systems can be generalized to more complex forms of matter, in particular to liquid crystals and polymers."*

# Pierre-Gilles de Gennes

**Gennes, Pierre-Gilles de** (1932- ), French physicist and winner of the 1991 Nobel Prize in physics for his work on polymers (molecular compounds characterized by long chains of repeating units).

Born in Paris, France, de Gennes completed his undergraduate work at the prestigious École Normale Supérieure in Paris, and received his Ph.D. degree in research science from the Centre d'Études Nucléaires de Saclay in 1959. In 1961 he accepted a position as professor of solid-state physics at the University of Paris, Orsay, a post he held for ten years. He became a professor at the Collège de France in Paris in 1971. In 1976 de Gennes was appointed director of the École de Physique et Chimie in Paris, and in 1988 he became the science director for chemical physics at Rhone-Poulenc, one of the largest chemical companies in France.

De Gennes analyzed the characteristics of polymers by comparing them to simpler systems like magnets and liquid crystals. Once he discovered that these simpler systems shared mathematical properties with the more complex polymers, he was able to show, for example, that the thickness of the polymer chain depends on its length. Understanding the fundamental nature of these complex substances has provided insight into how molecules are arranged in other substances ranging from super glue to liquid helium. This research also has helped scientists manipulate important properties of these substances to get the best results when designing new molecules.

De Gennes founded the STRASACOL, a joint project between physicists and chemists from Strasbourg, Saclay, and the Collège de France for the study of polymers. He is also the author of several books, including *The Physics of Liquid Crystals (1974), Superconductivity of Metals* and *Alloys (1966), and Scaling Concepts in Polymer Physics (1979),* considered classic texts in the field.

Georges Charpak

*"Polish-born French physicist Georges Charpak won the Nobel Prize in physics, in 1992. Charpak invented devices that were used to detect atomic particles."*

## GEORGES CHARPAK

**Charpak, Georges** (1924- ), Polish-born French physicist and winner of the 1992 Nobel Prize in physics for his invention of devices for detecting atomic particles, such as the multiwire proportional chamber. Charpak's inventions liberated scientists from dependence on film to detect and record subatomic matter, greatly advancing the study of nuclear processes.

Born in Dabrovica, Poland, Charpak emigrated to France with his family at the age of five. When he was 19, the French Vichy government (the government that headed France during its occupation by Nazi Germany during World War II) accused Charpak of terrorism. They shipped him to the concentration camp in Dachau, Germany, where he was incarcerated until liberation in 1945. Charpak returned to France to earn a degree in civil engineering at the École des Mines, Paris, and followed this with graduate studies in nuclear physics at the Collège de France, Paris. He became a French citizen during this time. In 1948 Charpak accepted a position as researcher in the laboratory of physicist Frédéric Joliot-Curie, where he began to work on designing particle detection equipment. In 1955 Charpak was awarded his Ph.D. degree in nuclear physics from the Collège de France, and in 1958 he accepted a research position at the European Organization of Nuclear Research (CERN, now the European Laboratory for Particle Physics). There, he concentrated his work on the study of particle detection. In 1984, while remaining affiliated with CERN, Charpak was named the Joliot-Curie Professor at the École Supérieure de Physique et Chimie in Paris.

The multiwire proportional chamber consists of a number of wires in a chamber of ionized gas. The wires attract electrons, and a computer analyzes the current produced in the wires as this happens. Charpak's multiwire invention replaced the slower and less efficient technique of photographic analysis of nuclear particles and greatly advanced the study of the nature of matter. In recent years, Charpak has focused his attention on the field of medicine, analyzing the structure of protein with superfast X rays and studying receptors in the brain.

Charpak was also awarded the High Energy and Particle Physics Prize from the European Physical Society (1989). Inspired by his own wrongful incarceration, Charpak is the founder of CERN's SOS Committee, which represents scientists whose civil rights have been denied by repressive governments. Among the scientists the committee has aided are the Soviet dissidents Andrey Sakharov and Natan (Anatoly) Sharansky. Charpak is a member of the French Academy of Sciences.

Joseph Taylor Jr.

*"American physicist Joseph Taylor, Jr., won the 1993 Nobel Prize in physics. His discovery of a new type of pulsar, or neutron star emitting energy pulses at regular intervals, allowed him new opportunities to study gravity and to test Einstein's theory of general relativity."*

## JOSEPH H. TAYLOR JR.

**Taylor, Joseph H., Jr.** (1941- ) American physicist and Nobel Prize winner. With colleague American physicist Russell A. Hulse, he discovered the first binary pulsar, which is a pair of rotating neutron stars. Taylor and Hulse used this discovery to study characteristics of gravity and also to test physicist Albert Einstein's general theory of relativity. For this work, Taylor and Hulse shared the 1993 Nobel Prize in physics.

Taylor studied radio astronomy at Harvard University, receiving his Ph.D. degree in 1968. That same year, he joined the faculty at the University of Massachusetts, Amherst, where he helped several area schools organize the Five College Radio Astronomy Observatory. In 1981 Taylor joined the faculty at Princeton University, where he continues to study binary pulsars.

Pulsars were first discovered in the 1960s, and since that time hundreds have been found. A pulsar is a type of neutron star, the collapsed core of a star after it explodes. A pulsar spins rapidly, is extremely dense, and is surrounded by especially strong magnetic fields. Beams of radiation are emitted from the rotating neutron star, sweeping out through space as the star spins. Astronomers detect these moving beams as short, regular pulses of radiation. In the early 1970s Taylor and Hulse undertook an extensive search for new pulsars, using the 305 m (1000 ft), bowl-shaped reflector of the Arecibo Observatory in Puerto Rico. By using computers to carefully process the signals received from space, the scientists were able to pinpoint the location of various new pulsars. In 1974 Hulse observed a pulsar whose characteristics were unlike those of any other known rotating neutron star. Pulsars are noted for the very high accuracy of their radiation pulses. The cycle of this newly discovered pulsar changed by about 45 minutes each day, which led Taylor and Hulse to conclude, correctly, that they had found a binary rotating neutron star-that is, a pair of pulsars spinning around each other. The strong magnetic fields of this binary system provided the researchers with a new opportunity to test and verify aspects of Einstein's general theory of relativity. Predictions about decay of the pulsar orbits and the existence of gravity waves seemed to be confirmed by observations of this binary pulsar. Since then, several dozen binary pulsars have been found and studied, further supporting Taylor and Hulse's findings.

Russell Hulse

*"American physicist Russell Hulse won the 1993 Nobel Prize in physics. He won the award for his discovery of a new kind of pulsar, which led to a greater understanding of the characteristics of gravity."*

# RUSSELL A. HULSE

**Hulse, Russell A.** (1950- ), American physicist and co-winner of the 1993 Nobel Prize, with colleague Joseph H. Taylor Jr., for their discovery of the first binary pulsar, a pair of neutron stars that orbit each other. A neutron star is the collapsed core of a star after it explodes. Hulse and Taylor used this discovery to study characteristics of gravity and also to test German American physicist Albert Einstein's general theory of relativity.

Hulse received his Ph.D. degree in 1975 from the University of Massachusetts, where he was a graduate student under Taylor. After postdoctoral work at the National Radio Astronomy Observatory in Virginia, Hulse changed his research focus from astronomy to plasma physics and joined, in 1977, the staff of the Princeton Plasma Physics Laboratory at Princeton University.

Pulsars were first discovered in the 1960s, and since that time hundreds have been found. A pulsar spins rapidly, is extremely dense, and is surrounded by tremendously powerful magnetic fields. Beams of radiation are emitted from the pulsar, sweeping out through space as the pulsar spins. Astronomers detect these moving beams as short, regular pulses of radiation. In the early 1970s, Hulse and Taylor undertook an extensive search for new pulsars using the 305-meter (1000 feet), bowl-shaped reflector of the Arecibo Observatory in Puerto Rico. By using computers to carefully process the signals received from space, they were able to pinpoint the locations of various new pulsars.

In 1974 Hulse observed a pulsar whose characteristics were unlike any other known pulsar. Hulse and Taylor correctly concluded that they had found a binary pulsar, a pair of pulsars orbiting each other. The strong gravitational field of the binary system and the acceleration of the pulsars as they orbited each other provided the researchers with a new opportunity to test and verify aspects of Einstein's general theory of relativity. Einstein predicted in his theory that accelerating bodies would radiate energy in the form of gravitational waves. Einstein's predictions about the existence of gravity waves seemed to be confirmed by observations that showed that this binary pulsar was indeed losing energy. Since then, several dozen binary pulsars have been found and studied, further supporting Hulse and Taylor's findings.

Clifford Shull

*"American physicist Clifford Shull won the 1994 Nobel Prize in physics He was awarded the prize for helping develop neutron scattering, a technique used to study atomic structure and magnetic properties of materials."*

# CLIFFORD G. SHULL

**Shull, Clifford G.** (1915-2001) American physicist and co-winner of the 1994 Nobel Prize in physics for the development of the technique of neutron scattering, which is used to study the atomic structure and magnetic properties of various materials. He shared the prize with Canadian physicist Bertram Neville Brockhouse.

Shull attended the Carnegie Institute of Technology and then New York University, where he received his Ph.D. degree in 1941. He worked in private industry for five years and then joined the staff at the Oak Ridge National Laboratory in Tennessee. In 1995 he joined the faculty of the Massachusetts Institute of Technology (MIT).

Shull began studying neutron scattering in the 1940s using neutrons produced from nuclear reactors. At the time, scientists knew that beams of neutrons are scattered, or diffracted, by simple crystals, such as sodium chloride (table salt). However, they did not yet realize the potential of these beams for providing information about the atomic structure of a substance. Because neutrons are uncharged, they pass readily into a substance, much like X rays, but when the neutrons strike atomic nuclei in the substance, they are scattered at various angles. By using this information, Shull was able to study the scattering characteristics of dozens of atomic nuclei. He found that different atoms produce different diffraction results. He also discovered that the results for an atom of a particular element are always the same, independent of the other elements to which the atom is bonded. For example, a sodium atom scatters neutrons in a certain way, whether it is found in sodium chloride or sodium bromide. This uniformity proved the usefulness of neutron scattering for studying the atomic structure of many different kinds of materials. In addition, the small magnetic properties of neutrons were helpful in studying the structure of magnetic materials.

Neutron scattering is now widely used in laboratories throughout the world to study and develop various materials, including polymers, semiconductors, and superconductors.

Bertram Brockhouse

*"Canadian physicist Bertram Brockhouse won the Noble Prize in physics in 1994. Brockhouse used neutron scattering to study atomic structure and movement."*

# BERTRAM N. BROCKHOUSE

**Brockhouse, Bertram N.** (1918- ) Canadian physicist and Nobel Prize winner. He helped to develop the technique of neutron scattering to study atomic structure and movement in various materials. He shared the 1994 Nobel Prize in physics with American physicist Clifford G. Shull for this work.

Brockhouse attended the University of British Columbia in Vancouver and received his Ph.D. degree from the University of Toronto in 1950. For more than a decade, he conducted nuclear-reactor research at the Chalk River Laboratories in Ontario. In 1962 he joined the faculty of McMaster University, also in Ontario, and remained there until retirement. He became professor emeritus in 1984.

Building on the work of Clifford Shull and other scientists who first developed neutron-scattering techniques, Brockhouse designed a new instrument for neutron-scattering research. Neutron scattering is based on the fact that beams of neutrons can pass readily into a substance, much as X rays can. The neutrons are scattered, or diffracted, by the atoms in the substance. By measuring the neutron change, scientists can gather information about a substance's atomic structure. Brockhouse designed an instrument called a triple-axis spectrometer, which can measure the energy and momentum of the neutrons as they enter and leave the sample substance. This allowed him to gather data about vibrations and other movements of the atoms within the substance, ultimately providing information about the substance's physical properties. His instrument is still widely used in neutron-scattering studies, which have applications in biology, chemistry, materials science, and engineering. Polymers, semiconductors, and superconductors are just three of the many materials analyzed and developed with the help of neutron-scattering techniques. Brockhouse's work is also credited with helping to form the basis of modern solid-state physics, also known as condensed-matter physics.

Martin Perl

*"American physicist Martin Perl won the Nobel Prize in physics in 1995 for the discovery of a new subatomic particle, the tau lepton."*

# MARTIN L. PERL

**Perl, Martin L.** (1927- ), American physicist and co-winner of the 1995 Nobel Prize in physics for his discovery of the tau lepton, a type of fundamental particle that makes up part of the atom. The 1995 prize was divided between Perl and American physicist Frederick Reines.

Perl, born in New York City, studied chemical engineering as an undergraduate at the Brooklyn Polytechnic Institute in New York. He received his Ph.D. degree in physics from Columbia University in 1950. Perl taught at the University of Michigan and then in 1964 joined the faculty of Stanford University, where he worked at the Stanford Linear Accelerator Center (SLAC).

The lepton is one of three groups of elementary particles, along with quarks and bosons. There are six known types of leptons, three that carry a negative electric charge and three with no charge. Of the negatively charged leptons, the electron was discovered in 1897, and the muon-basically a heavy electron-was discovered in the 1940s. Uncharged leptons are called neutrinos. Before Perl's discovery, theoretical physicists had not even predicted the existence of the tau, the third charged lepton. In 1973, Perl set out to study electrons and muons and to look for any differences between the two particles. By colliding subatomic particles, Perl was able to study both the particles and the products they produced after the collision. Occasionally, a collision resulted in the production of electrons and muons. Since Perl knew that muons were unstable particles that decayed, or broke down, into electrons and neutrinos, he theorized that only the decay of a more massive particle could result in the production of muons. In 1975, after analyzing tens of thousands of pieces of data, Perl announced his discovery of the tau lepton, a particle that is about 3500 times heavier than the electron.

Frederick Reines

*"American physicist Frederick Reines won the Nobel Prize in physics in 1995 for his work confirming the existence of the neutrino, a subatomic particle."*

## FREDERICK REINES

**Reines, Frederick** (1918-1998) American physicist and co-winner of the 1995 Nobel Prize in physics for his confirmation of the existence of the neutrino, a type of subatomic particle. The 1995 prize was divided between Reines and American physicist Martin L. Perl.

Reines earned his Ph.D. at New York University in 1944, then joined the Manhattan Project in Los Alamos, New Mexico, helping to produce the atomic bomb. He remained at Los Alamos until 1959, when he joined the faculty of Case Western Reserve University in Ohio. In 1966 he took a position at the University of California at Irvine.

Neutrinos were first described in 1930 by Wolfgang Pauli, winner of the 1945 Nobel Prize in physics. In his studies of radioactivity, Pauli found that during the decay of a radioactive atomic nucleus, the energy of the electrons emitted did not account for all of the energy loss that he observed. He invented the concept of a new, uncharged subatomic particle that also was emitted from the nucleus during its decay. His reasoning was based on a fundamental law of physics-the law of conservation of energy. This law states that the amount of energy in the universe remains constant-it cannot be created or destroyed. Reines believed that this law held true even at the atomic and subatomic level. In the 1950s, Reines and his colleague, physicist Clyde L. Cowan, Jr., designed an experiment involving the collision of atomic particles in water. After almost a year, Reines and Cowan had collected enough evidence to conclusively announce in 1956 that neutrinos do exist as free particles. As Pauli had suspected, the neutrinos are emitted with electrons during radioactive decay.

David M. Lee

*"David M. Lee American physicist and Nobel laureate. Lee helped discover that a rare form of helium, known as helium-3, exhibits the phenomenon of superfluidity at extremely low temperatures. He won the Nobel Prize in physics in 1996."*

# DAVID M. LEE

**Lee, David M.** (1931- ), American physicist and Nobel laureate. Lee helped discover that a rare form of helium, known as helium-3, exhibits the phenomenon of superfluidity at extremely low temperatures. In a superfluid state, atoms move together in such a way that the fluid flows with no resistance. A superfluid may climb the walls of a container and may even flow uphill. The discovery made by Lee and his colleagues in the early 1970s launched vigorous investigations into this form of helium and its superfluid properties.

Lee was born in Rye, New York. He earned his bachelor's degree from Harvard University in 1952 and his master's degree in 1955 from the University of Connecticut. He obtained his doctoral degree in physics from Yale University in 1959. That same year he joined the physics faculty at Cornell University in Ithaca, New York.

Superfluidity was discovered independently by two scientists in 1937 and in 1938, in a more abundant form of helium known as helium-4. Helium-4 atoms contain one more neutron in their nucleus than helium-3 atoms. For many years afterward, scientists believed that these two forms, or isotopes, of helium were so dissimilar in their atomic structures that helium-3 would not exhibit superfluidity. In the late 1950s, however, physicists developed newer theories that suggested that helium-3 would become a superfluid at a very low temperature. Many labs attempted to achieve this transition, but without success.

Lee began working with helium-3 with a colleague, physicist Robert C. Richardson, and a graduate student, Douglas D. Osheroff. The three scientists were not looking for superfluidity, but were cooling helium-3 to extremely low temperatures-within a few thousandths of a degree of absolute zero (approximately -273.16° C/-459.69° F)-in order to explore the magnetic properties of the isotope.

While working in the lab in 1971, Osheroff noticed indications of pressure changes within the helium-3. He made special note of the changes, believing that they indicated the helium-3 had undergone the transition to a superfluid. The three scientists soon learned that Osheroff was correct, but their results were so unexpected that the scientific community did not accept them until other teams reproduced the results over the next year.

A liquid in the superfluid state does not behave according to the same laws of physics that normal fluids, such as water, obey. Instead, superfluids are subject to the complex statistical rules of quantum mechanics, the branch of physics that predicts how atomic and subatomic particles behave. Superfluids are important to physicists because they offer a macroscopic, or relatively large, system in which scientists can study quantum effects. For their discovery of superfluidity in helium-3, Lee, Richardson, and Osheroff shared the 1996 Nobel Prize in physics.

Douglas D. Osheroff

*"Douglas D. Osheroff American physicist and Nobel laureate. Osheroff helped discover that a rare form of helium known as helium-3, exhibits the phenomenon of superfluidity at extremely low temperatures. He won the Nobel Prize in physics in 1996."*

# Douglas D. Osheroff

**Osheroff, Douglas D.** (1945- ), American physicist and Nobel laureate. Osheroff helped discover that a rare form of helium, known as helium-3, exhibits the phenomenon of superfluidity at extremely low temperatures. In a superfluid state, atoms move together in such a way that the fluid flows with no resistance. A superfluid may climb the walls of a container and may even flow uphill. The discovery made by Osheroff and his colleagues in the early 1970s launched vigorous investigations into this form of helium and its superfluid properties.

Osheroff was born in Aberdeen, Washington. He earned his bachelor's degree from the California Institute of Technology in 1967 and his doctoral degree from Cornell University in 1973. Osheroff did research at the American Telephone and Telegraph (AT&T) Bell Laboratories from 1972 to 1987, when he became a professor of physics and applied physics at Stanford University.

Superfluidity was discovered independently by two scientists in 1937 and in 1938, in an abundant form of helium known as helium-4. Helium-4 atoms contain one more neutron in their nucleus than helium-3 atoms. For many years afterward, scientists believed that these two forms, or isotopes, of helium were so dissimilar in their atomic structures that helium-3 would not exhibit superfluidity. In the late 1950s, however, physicists developed newer theories that suggested that helium-3 would become a superfluid at a very low temperature. Many labs attempted to achieve this transition, but without success.

Osheroff began working with helium-3 as a graduate student at Cornell with his adviser, physicist David M. Lee, and another faculty member, physicist Robert C. Richardson. The three scientists were not looking for superfluidity, but were cooling helium-3 to extremely low temperatures-within a few thousandths of a degree of absolute zero (approximately -273.16° C/-459.69° F)-in order to explore the magnetic properties of the isotope.

While working in the lab in 1971, Osheroff noticed indications of pressure changes within the helium-3. He made careful note of the changes, believing that they indicated the helium-3 had undergone the transition to a superfluid. The three scientists soon learned that Osheroff was correct, but their results were so unexpected that the scientific community did not accept them until other teams reproduced the results over the next year.

A liquid in the superfluid state does not behave according to the same laws of physics that normal fluids, such as water, obey. Instead, superfluids are subject to the complex statistical rules of quantum mechanics, the branch of physics that predicts how atomic and subatomic particles behave. Superfluids are important to physicists because they offer a macroscopic, or relatively large, system in which scientists can study quantum effects. For their discovery of superfluidity in helium-3, Osheroff, Lee, and Richardson shared the 1996 Nobel Prize in physics.

Robert C. Richardson

*"Robert C. Richardson American physicist and Nobel laureate. Richardson helped discover that a rare form of helium, known as helium-3, exhibits the phenomenon of superfluidity at extremely low temperatures. He won the Nobel Prize in physics in 1996."*

# ROBERT C. RICHARDSON

**Richardson, Robert C.** (1937- ),American physicist and Nobel laureate. Richardson helped discover that a rare form of helium, known as helium-3, exhibits the phenomenon of superfluidity at extremely low temperatures. In a superfluid state, atoms move together in such a way that the fluid flows with no resistance. A superfluid may climb the walls of a container and may even flow uphill. The discovery made by Richardson and his colleagues in the early 1970s launched vigorous investigations into this form of helium and its superfluid properties.

Richardson was born in Washington, D.C. He studied physics at the Virginia Polytechnic Institute, receiving his bachelor's degree in 1958 and his master's degree in 1960. He earned his Ph.D. degree in physics from Duke University in 1966. That same year he became a research assistant at Cornell University in Ithaca, New York. By 1975 he had become a full professor at Cornell. In 1990 he became director of the university's Laboratory of Atomic and Solid State Physics.

Superfluidity was discovered independently by two scientists in 1937 and in 1938, in an abundant form of helium known as helium-4. Helium-4 atoms contain one more neutron in their nucleus than helium-3 atoms. For many years afterward, scientists believed that these two forms, or isotopes, of helium were so dissimilar in their atomic structure that helium-3 would not exhibit superfluidity. In the late 1950s, however, physicists developed newer theories that suggested that helium-3 would become a superfluid at a very low temperature. Many labs attempted to achieve this transition, without success.

Richardson began working with helium-3 with a colleague, physicist David M. Lee, and a graduate student, Douglas D. Osheroff. The three scientists were not looking for superfluidity, but were cooling helium-3 to extremely low temperatures-within a few thousandths of a degree of absolute zero (approximately -273.16° C/-459.69° F)-in order to explore the magnetic properties of the isotope.

While working in the lab in 1971, Osheroff noticed indications of pressure changes within the helium-3. He made careful note of the changes, believing that they indicated the helium-3 had undergone the transition to a superfluid. The three scientists soon learned that Osheroff was correct, but their results were so unexpected that the scientific community did not accept them until other teams reproduced the results over the next year.

A liquid in the superfluid state does not behave according to the same laws of physics that normal fluids, such as water, obey. Instead, superfluids are subject to the complex statistical rules of quantum mechanics, the branch of physics that predicts how atomic and subatomic particles behave. Superfluids are important to physicists because they offer a macroscopic, or relatively large, system in which scientists can study quantum effects. For their discovery of superfluidity in helium-3, Richardson, Lee, and Osheroff shared the 1996 Nobel Prize in physics.

Steven Chu

*"American physicist Steven Chu shared the 1997 Nobel Prize in physics with two other researchers. Chu was honoured for his work in cooling and trapping atoms and other small particles."*

# STEVEN CHU

**Chu, Steven** (1948- ), American physicist and Nobel laureate. Chu led a team of physicists in the mid-1980s who were the first to trap atoms (one of the basic units of matter) with special beams of light called lasers. These laser traps allow scientists to study atoms more closely and use atoms more efficiently in devices such as atomic clocks. Chu's research helped pave the way for important discoveries in atomic physics. It also led to the development of many new practical applications, including more accurate atomic clocks and more precise devices for measuring the pull of gravity. He shared the 1997 Nobel Prize for physics with two other researchers who made separate and complementary advancements in the field, French physicist Claude Cohen-Tannoudji and American physicist William D. Phillips.

Chu was born in St. Louis, Missouri, and grew up on Long Island, New York. He earned dual undergraduate degrees in physics and mathematics at the University of Rochester in New York in 1970. In 1976 he received his doctoral degree in physics from the University of California, Berkeley. Chu spent two years conducting post-doctoral research at Berkeley, then served as a member of the technical staff of American Telephone and Telegraph (AT&T) from 1978 to 1983.

In 1983 Chu was named head of the quantum electronics research department at AT&T's Bell Laboratories and moved to the lab's complex in Holmdel, New Jersey. That year he began discussing the possibility of trapping atoms with American physicist Arthur Ashkin. Over the next few years, Chu, Ashkin, and fellow American physicists John Bjorkholm and Alex Cable conducted experiments in which six lasers bombarded atoms from six sides in a vacuum chamber at Chu's lab.

At room temperature, atoms move at speeds of about 4000 km/h (2500 mph), much too fast for scientists to easily study them. The speed of atoms is related to their temperature-atoms at a higher temperature move faster than cooler atoms. Slowing a sample of atoms will therefore make the atoms cooler. Chu pioneered a technique for slowing and cooling atoms that uses lasers to immerse the atoms in photons (packets of light wave energy). The photons strike the atoms in a way that is roughly analogous to raindrops hitting a beach ball. The photons have no mass, but because they move at the speed of light, they carry momentum and can affect a small mass, such as an atom. In Chu's trap, the impact of enough photons hitting atoms slowed the atoms down. The atoms moved so slowly that they seemed to be stuck, so Chu's team named the process "optical molasses."

In 1985 Chu and his team cooled atoms to 240 millionths of a Celsius degree (430 millionths of a Fahrenheit degree) above absolute zero, the point at which all matter stops moving (-273.15° C, or -459.67° F). By 1988 William D. Phillips and his research team had cooled atoms to 40 millionths of a Celsius degree (70 millionths of a Fahrenheit degree) above absolute zero, which was lower than scientists believed to be theoretically possible at the time. Claude Cohen-Tannoudji and his team achieved a temperature of 0.2 millionths of a Celsius degree (0.4 millionths of a Fahrenheit degree) above absolute zero in 1995.

The ability to manipulate atoms and study them more closely led to many real and potential applications, including increased accuracy in atomic clocks, which improved their use in space navigation and global positioning systems. Atomic clocks keep track of time by counting waves of radiation emitted by special atoms in traps inside the clock. If the traps can hold the atoms at lower temperatures, the traps and mechanisms inside the clock can exercise more control over the atom, reducing the possibility of error. Atom manipulation also contributes to increased accuracy of the measurement of gravitational force, which is useful in, among other things, oil exploration. This is because a deposit of oil or other substance beneath the earth's crust has a different density from the areas around it. Changes in density produce changes in the local gravitational field, because the gravitational force of an area is related to the mass of that area. Advances in the manipulation of atoms have also raised the possibility of using atoms to etch electronic circuits, thereby increasing the circuits' capabilities.

In 1987 Chu joined the faculty at Stanford University as a professor of physics and applied physics. While at Stanford, he and his colleagues continued the study of cooled atoms. In 1992 Chu was named a fellow of the American Academy of Arts and Sciences and a year later a member of the National Academy of Sciences.

Claude Cohen-Tannoudji

*"French physicist Claude Cohen-Tannoudji shared the 1997 Nobel Prize in physics with two other physicists. All three physicists were recognized for their individual work in cooling and trapping atoms."*

# Claude Cohen-Tannoudji

**Cohen-Tannoudji, Claude** (1933-), French physicist and Nobel laureate. Cohen-Tannoudji pioneered research into cooling, slowing, and trapping atoms-one of the basic building blocks of matter-with special beams of light called lasers. The techniques Cohen-Tannoudji and others developed led to significant advancements in the study and manipulation of atoms, resulting in many applications. These applications include more accurate atomic clocks and more precise devices for measuring gravity. He shared the 1997 Nobel Prize for physics with two American scientists who were also successful in slowing atoms, Steven Chu and William D. Phillips. The three didn't work side by side; yet each contributed to and built upon the work of the other two.

Born in Constantine, Algeria, Cohen-Tannoudji earned his undergraduate degree in 1957 and his Ph.D. degree in 1962 in physics from École Normale Supérieure (ENS) in Paris, France. As a student at ENS he worked under the direction of French physicist Alfred Kastler, who won the Nobel Prize in physics in 1966, and Jean Brossel, another noted French physicist. Cohen-Tannoudji remained a researcher at ENS throughout his career. He was instrumental in the creation of a research center devoted to atomic physics and optics. The center is named the Kastler-Brossel Laboratory, after his two professors.

Cohen-Tannoudji joined the Centre National de la Recherche Scientifique (CNRS) in 1960. In 1973, while still associated with CNRS, he began a lengthy stint as chair of atomic and molecular physics at the Collège de France in Paris.

As a student, researcher, and professor, Cohen-Tannoudji became an expert in the field of slowing atoms. At room temperature, atoms move at speeds of about 4000 km/h (about 2500 mph), too fast for scientists to study them thoroughly. Heat results from atomic motion, so slowing atoms down lowers their temperature. Cohen-Tannoudji was among the first to propose using lasers to slow down atoms. The concept involves bombarding the atoms with laser light. Packets of light wave energy called photons strike the atoms in a way that is roughly the same as raindrops hitting a beach ball. Even though the photons have no mass, they move fast enough (at the speed of light) to produce enough momentum to slow the atoms with their impacts.

In 1985 Steven Chu and his team cooled atoms to 240 millionths of a Celsius degree (430 millionths of a Fahrenheit degree) above absolute zero (-273.15° C, or -459.67° F), which slowed atoms to about 0.5 km/h (0.3 mph). By 1988 William D. Phillips and his research team had cooled atoms to 40 millionths of a Celsius degree (70 millionths of a Fahrenheit degree) above absolute zero, which was below the temperature that was believed to be theoretically possible at the time. In 1995 Cohen-Tannoudji and his team achieved a temperature of 0.2 millionths of a Celsius degree (0.4 millionths of a Fahrenheit degree) above absolute zero. Through his career Cohen-Tannoudji developed several cooling mechanisms used to trap atoms.

Manipulating atoms and studying them more closely allows scientists to improve many areas of technology. In an atomic clock, the more control the mechanism has over the atoms that it uses to keep track of time, the more accurate it can be. Increasingly accurate atomic clocks improve space navigation and global positioning systems that rely on the clocks. Trapped atoms are also useful in measuring the force of gravity of a particular spot on the earth. Changes in the gravitational field of the earth from place to place indicate changes in the density of the earth, which can lead scientists to oil and other valuable deposits within the earth. The ability to control atoms also raises the possibility of using atoms to etch electronic circuits. This would increase the circuits' capabilities by making the circuits finer, and each area of circuit board would be able to hold more circuits.

In 1996 Cohen-Tannoudji won CNRS's gold medal. In 1997, along with former students Jacques Dupont-Roc and Gilbert Grynberg, he wrote the book *Introduction to Quantum Electrodynamics*.

William Phillips

*"American physicist William Phillips shared the 1997 Nobel Prize in physics with two other researchers. Phillips was recognized for his work on ways to cool and trap atoms."*

# William D. Phillips

**Phillips, William D.** (1948-), American physicist and Nobel laureate. Phillips's advancements in the use of special beams of light called lasers to slow, cool, and capture atoms (tiny particles that make up matter) were instrumental in furthering the study and use of atoms. In the late 1980s Phillips used laser cooling to cool and slow atoms to a point not thought possible at the time. He shared the 1997 Nobel Prize for physics with two other scientists who made separate but complementary advancements, Steven Chu of the United States and Claude Cohen-Tannoudji of France. Their achievements led to a breakthrough in the study and manipulation of atoms, which in turn brought improvements to many applications, including global navigation and gravitational measurement techniques.

Phillips was born in Wilkes-Barre, Pennsylvania. He earned a B.S. degree in physics at Juniata College in Huntingdon, Pennsylvania, in 1970. In 1976 he earned his Ph.D. degree in physics from Massachusetts Institute of Technology (MIT). After post-doctoral research at MIT, in 1978 he joined the National Institute of Standards and Technology (NIST), then known as the National Bureau of Standards.

Hired to work with precision electrical measurements, Phillips soon also began conducting experiments in trapping atoms. He made advancements using a magnetic device to slow atoms. Meanwhile Steven Chu and a team at Bell Laboratories in Holmdel, New Jersey, began furthering the use of lasers to capture atoms. In 1985 Chu successfully used lasers in a vacuum chamber to cool atoms to 240 millionths of a Celsius degree (430 millionths of a Fahrenheit degree) above absolute zero, the point at which all matter stops moving (-273.15° C, or -459.67° F).

Phillips adopted Chu's techniques, and by 1988 Phillips and his research team had cooled atoms to 40 millionths of a Celsius degree (70 millionths of a Fahrenheit degree) above absolute zero, lower than scientists thought was theoretically possible at the time. Phillips came up with methods to capture atoms at regular intervals in what was termed an optical lattice.

At room temperature, atoms move at speeds of about 4000 km/h (2500 mph), much too fast for scientists to study them. The rate at which atoms move is related to the temperature of the matter made up by the atoms. Lowering the temperature of the sample of atoms slows the atoms' motion, and vice versa. Chu and Phillips developed techniques in which atoms are bombarded with finely tuned laser beams. The lasers immerse the atoms in packets of light wave energy called photons. The photons strike the atoms in a way that is roughly like raindrops hitting a beach ball. The photons have no mass, but because they travel at the speed of light, they carry enough momentum to hit the atoms and slow them down. By 1995 Claude Cohen-Tannoudji and his team used similar techniques to lower the temperature of a sample of atoms to 0.2 millionths of a Celsius degree (0.4 millionths of a Fahrenheit degree) above absolute zero.

The ability to manipulate atoms and study them more closely led to many immediate and many potential applications. Trapped atoms have increased the accuracy of atomic clocks, which increases the accuracy of other instruments that use atomic clocks, such as navigation systems. Control of atoms also helps calibrate instruments used to measure the force of gravity at spots on the earth. These measurements indicate different densities within the earth, which can reveal features such as petroleum deposits beneath the earth's surface. The ability to manipulate atoms also raises the possibility of using atoms to etch electronic circuits, thereby increasing the circuits' capabilities by increasing the number of circuits that can fit in a certain area.

In the 1990s Phillips continued his research into ultra-cold trapped atoms. In 1995 he was elected to the American Academy of Arts and Sciences and became an NIST Fellow. Two years later he was named to the National Academy of Sciences.

Robert Laughlin

*"American physicist Robert Laughlin shared the 1998 Nobel Prize in physics with two other researchers. Laughlin was honoured for his theoretical work on the fractional quantum Hall effect. The fractional quantum Hall effect occurs when electrons act together to form particles in which the electrons appear to have fractions of their normal electric charge."*

# ROBERT B. LAUGHLIN

**Laughlin, Robert B.** (1950- ), American physicist and Nobel Prize winner. Laughlin shared the 1998 Noble Prize in physics with Chinese-born American physicist Daniel Tsui and German-born American physicist Horst Stormer. The three men collaborated on the discovery that electrons (tiny, negatively charged particles) can act together to form particle-like units called quasiparticles. When electrons form these quasiparticles, they appear to have only a fraction of their normal electric charge. Laughlin provided the theoretical analysis to explain Stormer and Tsui's experimental discovery of this phenomenon, called the fractional quantum Hall effect.

Laughlin was born in Visalia, California. He graduated with a bachelor's degree in physics from the University of California at Berkeley in 1972. He continued his studies of physics at the Massachusetts Institute of Technology in Cambridge, Massachusetts, where he received his doctorate in physics in 1979.

In 1979 Laughlin went to work at AT&T's Bell Laboratories, now part of Lucent Technologies, in New Jersey. In 1982 he became a research physicist at Lawrence Livermore National Laboratory in California. He became an associate professor of physics at Stanford University in California in 1985 and a professor of physics at Stanford in 1989.

Laughlin did the work for which he won the Nobel Prize at Bell Labs in the early 1980s. Stormer and Tsui also worked at Bells Labs at that time. Stormer and Tsui discovered that electrons in certain conditions behave very strangely, appearing to have only a fraction of the electrical charge of individual electrons.

The experiment Stormer and Tsui performed was based on the Hall effect, a phenomenon discovered in 1879 by American physicist Edwin Hall. The Hall effect occurs when scientists place a conductor (a material that allows electricity to flow through it easily) carrying an electric current in a magnetic field. A magnetic field is a region of space surrounding a magnet that exerts a force on electrically charged or magnetized objects. The magnetic field in the Hall effect pushes electrons in the conductor off to one side of the conductor, creating an electric charge difference from one side of the material to the other. The current flows through the conductor, but a charge difference, or voltage, occurs from one side of the conductor to the other. For example, if the conductor is in the shape of a long, narrow rectangle and the current flows from one short end to the other, the voltage will occur from one long end to the other.

Stormer and Tsui studied conductors at very low temperatures. The electric characteristics, including the Hall effect, of materials change at low temperatures. In 1980 German physicist Klaus-Olaf von Klitzing examined the behavior of conductors at temperatures near absolute zero (-273° C, or -459° F). Von Klitzing also used very strong magnetic fields. He discovered that, in those conditions, the voltage created by the Hall effect does not change smoothly but varies in discrete steps. Von Klitzing won the 1985 Nobel Prize in physics for this discovery, called the integer quantum Hall effect, or quantized Hall effect.

Quantum theory, an area of physics that describes both electrons and radiation in terms of both waves and particles, explains these steps. Quantum theory states that electrons in a magnetic field move in circular orbits perpendicular, or at right angles, to the magnetic field. These orbits are also called energy levels. The number of electrons that each orbit can hold depends on the strength of the magnetic field-the stronger the magnetic field, the more electrons can share a single energy level. As von Klitzing gradually increased the strength of the magnetic field, the electrons would begin to collect at one side of the conductor. When the first energy level was full, the voltage would level off, then drop suddenly as electrons started filling up another energy level. Each new energy level created a new step in voltage. Each step was made up of a whole number of electrons, so the voltage difference between steps was always greater than or equal to the charge of an electron.

In 1982 Stormer and Tsui were testing the integer quantum Hall effect, but at temperatures even lower and in magnetic fields even stronger than those used by von Klitzing. They expected to see steps of charge equal to the electrical charge of a whole electron, as von Klitzing had discovered. Instead, they found that the steps were more numerous and smaller than the integer quantum Hall effect predicts. Stormer and Tsui knew that they had made an important discovery, but they could not explain what had happened. Electrons cannot split into smaller particles, and they could find no reason that the steps would be fractions of the charge of an electron.

Laughlin saw Stormer and Tsui's results and set out to explain them using quantum mechanical equations, the rules that describe the motion of subatomic particles. In 1983 Laughlin developed a new theory that involved groups of electrons acting together as a unit. He called these units quasiparticles. Laughlin's theory explains that each electron can belong to more than one quasiparticle and divide its charge between the quasiparticles. This allows for fractions of the charge of an electron to appear. In 1997 two groups of physicists saw direct evidence of quasiparticles. The effect, known as the fractional quantum Hall effect, does not have any immediate practical applications but may allow physicists to understand more about how the electrons in the early universe behaved. It also shows that the laws of quantum mechanics are powerful enough to explain new phenomena.

Laughlin continued his work in physics by studying high-temperature superconductivity when he moved to Stanford University in 1985. Superconductivity is a phenomenon that occurs in special materials below certain temperatures; electricity flows without resistance in these materials. Most superconductivity only occurs at very low temperatures. Laughlin's work dealt with superconductivity that occurs at relatively high temperatures for superconductors, far above absolute zero. He also continued to develop mathematical models using the fractional quantum Hall effect.

Daniel Tsui

*"Chinese-born American physicist Daniel Tsui shared the 1998 Nobel Prize in physics with two other physicists. Tsui was honoured for his experimental work on the fractional quantum Hall effect. The fractional quantum Hall effect occurs when electrons act together to form particles in which the electrons appear to have fractions of their normal electric charge."*

# DANIEL C. TSUI

**Tsui, Daniel C.** (1939- ), Chinese-born American physicist and Nobel Prize winner. Tsui shared the 1998 Nobel Prize in physics with American physicist Robert Laughlin and German-born American physicist Horst Stormer. These scientists discovered that electrons (tiny, negatively charged particles) in very strong magnetic fields can act together to form new types of particle-like units called quasiparticles. When electrons form these quasiparticles, they appear to have only a fraction of their normal electric charge. The phenomenon is called the fractional quantum Hall effect. Tsui and Stormer discovered the effect experimentally, and Laughlin provided the theoretical explanation for the effect.

Tsui was born in Henan, China. He came to the United States in 1958 to attend Augustana College in Rock Island, Illinois. He graduated from Augustana in 1961 with a bachelor's degree in mathematics. He received his doctorate in physics from the University of Chicago in 1967 and remained there as a research associate until 1968. In 1968 he joined AT&T's Bell Laboratories (now a part of Lucent Technologies) in New Jersey as a member of the technical staff. In 1982 he became a professor of electrical engineering at Princeton University in New Jersey.

Tsui and his colleague Horst Stormer did their Nobel Prize winning work in the early 1980s at Bell Laboratories. Their experiment was based on a phenomenon called the Hall effect, discovered by American physicist Edwin Hall in 1879. The Hall effect occurs when a conductor (a material that allows electricity to flow through it easily) carrying an electric current is placed in a magnetic field. A magnetic field is a region of space, influenced by a magnet, that exerts a force on charged or magnetic objects. The magnetic field pushes electrons in the conductor off to one side of the conductor, creating an electric charge difference from one side of the material to the other. The current flows through the conductor, but a charge difference occurs from one side of the conductor to the other. A charge difference in a material is called a voltage. For example, if the conductor is the shape of a long, narrow rectangle and the current flows from one short end to the other, the voltage difference will occur from one long end to the other.

Tsui and Stormer studied the Hall effect in materials at very low temperatures. The electrical characteristics, including the Hall effect, of a material change at low temperatures. In 1980 German physicist Klaus-Olaf von Klitzing examined the behavior of conductors at temperatures near absolute zero (-273° C, or -459° F). Von Klitzing also used very strong magnetic fields. He discovered that, under these conditions, the voltage created by the Hall effect does not change smoothly but varies in discrete steps. Von Klitzing won the 1985 Nobel Prize in physics for this discovery, called the integer quantum Hall effect, or quantized Hall effect.

Quantum theory, an area of physics that describes both electrons and radiation in terms of both waves and particles, explains these steps in the Hall effect. Quantum theory states that the electrons in a magnetic field move in circular orbits perpendicular to, or at right angles to, the magnetic field. These orbits are also called energy levels. The number of electrons that each orbit can hold depends on the strength of the magnetic field-the stronger the magnetic field, the more electrons can share a single energy level. As von Klitzing gradually increased the magnetic field, the electrons began to collect at one side of the conductor. When the first energy level was full, the voltage leveled off, then dropped suddenly. It began to rise again as electrons started filling up the next energy level. Each new energy level created a new step in voltage. Each step was made up of a whole number of electrons, so the voltage difference between steps was always greater than or equal to the charge of an electron.

In 1982 Tsui and Stormer tested the Hall effect by applying even stronger magnetic fields to material at even lower temperatures than those that had revealed the integer quantum Hall effect. They found that the voltage varied in steps, but the steps were far smaller than those in the integer quantum Hall effect. The steps were only fractions of the charge of an electron. Electrons cannot be split into smaller particles, so Tsui and Stormer could not explain their findings.

Robert Laughlin also worked at Bell Labs at the time and developed a theoretical explanation of the effect that Tsui and Stormer found. Laughlin developed the idea of quasiparticles, groups of electrons acting as a unit. Individual electrons can couple to more than one of these units, giving the appearance that the electron has a fraction of a whole charge. Quasiparticles are not new particles but rather electrons behaving together according to the laws of quantum mechanics. In 1997 two groups of physicists confirmed this theory by finding direct evidence of quasiparticles.

The effect that Tsui and Stormer discovered and Laughlin explained is known as the fractional quantum Hall effect. It does not have any immediate practical applications, but it will allow physicists to understand more about the behavior of electrons in the early universe. It could also play a role in the development of memory for computers by allowing physicists to use the fractional charges of quasiparticles instead of the whole number values of electron charge.

Horst Stormer

*"German-born American physicist Horst Stormer shared the 1998 Nobel Prize in physics with two other researchers. Stormer was honoured for his experimental work on the fractional quantum Hall effect. The fractional quantum Hall effect occur when electrons act together to form particles in which the electrons appear to have fractions of their normal electric charge."*

# HORST L. STORMER

**Stormer, Horst L.** (1949- ), German-born American physicist and Nobel Prize winner. Stormer shared the 1998 Nobel Prize in physics with American physicist Robert Laughlin and Chinese-born American physicist Daniel Tsui. Stormer and Tsui discovered that electrons (tiny particles with negative electric charges) in very strong magnetic fields can act together to form new types of particle-like units called quasiparticles. When electrons form these quasiparticles, they appear to have only a fraction of their normal electric charge. Laughlin explained the phenomenon.

Stormer was born in Frankfurt, Germany. He started college studying architecture but soon turned to physics. He received his undergraduate degree in physics from Johann Wolfgang Goethe University in Frankfurt in 1970 and his master's degree in physics from the same institution in 1974. He earned his Ph.D. degree in physics from Stuttgart University in 1977. In the same year, Stormer went to AT&T's Bell Laboratories, now part of Lucent Technologies, in New Jersey as a postdoctoral researcher. He joined Bell Labs as a member of the technical staff a year later. Stormer became head of the Semiconductor Physics Research Department in 1983.

Stormer served as director of the Physical Research Lab at Bell Laboratories from 1992 until 1997. In 1997 he returned to semiconductor research and became adjunct director of physics at Bell Laboratories. In 1998 he took an additional position as professor of physics and applied physics at Columbia University in New York.

Stormer did the work for which he won the Nobel Prize at Bell Laboratories in the early 1980s with Daniel Tsui. Their experiment was based on a phenomenon called the Hall effect, discovered by American physicist Edwin Hall in 1879. The Hall effect occurs when a conductor (a material that allows electricity to flow through it easily) carrying an electric current is placed in a magnetic field. A magnetic field is a region of space surrounding a magnet that exerts a force on electrically charged or magnetic materials. The magnetic field pushes electrons in the conductor off to one side of the conductor, creating an electric charge difference from one side of the material to the other. A charge difference in a material is called a voltage. The current flows through the conductor, but a voltage occurs from one side of the conductor to the other. For example, if the conductor is in the shape of a long, narrow, rectangle and the current flows from one short end to the other, the voltage will occur from one long edge to the other.

Stormer and Tsui studied conductors at very low temperatures. The electric characteristics, including the Hall effect, of materials change at low temperatures. In 1980 German physicist Klaus-Olaf von Klitzing examined the behavior of conductors at temperatures near absolute zero (-273°C, or -459° F). Von Klitzing also used very strong magnetic fields. He discovered that, in those conditions, the voltage created by the Hall effect varies in discrete steps. Von Klitzing won the 1985 Nobel Prize in physics for this discovery, called the integer quantum Hall effect, or quantized Hall effect.

Quantum theory, an area of physics that describes both particles, such as electrons, and radiation in terms of both waves and particles, explains these steps in voltage. Electrons are tiny negatively charged particles that create the charge difference in the Hall effect. Quantum theory states that electrons in a magnetic field move in circular orbits perpendicular to, or at right angles to, the magnetic field. These orbits are also called energy levels. The number of electrons that each orbit, or energy level, can hold depends on the strength of the magnetic field-the stronger the magnetic field, the more electrons can share a single energy level. As von Klitzing gradually increased the magnetic field, the electrons began to collect at one side of the conductor. When the first energy level was full, the voltage leveled off, then dropped suddenly as electrons began filling up another energy level. Each new energy level created a new step in voltage. Each step was made up of a whole number of electrons, so the voltage difference between steps was always greater than or equal to the charge of an electron.

In 1982 Stormer and Tsui tested the quantum Hall effect but used even stronger magnetic fields at even colder temperatures than those used by von Klitzing. Stormer and Tsui expected to find steps of charge in increments equal to the charge of an electron, as von Klitzing had discovered. Instead, they found many more steps that were smaller than they expected. These steps were only a fraction of the charge of an electron. Electrons cannot split into smaller particles, so Stormer and Tsui could not explain how the steps could be less than the charge of a whole electron.

Their colleague at Bell Laboratories, Robert Laughlin, saw that they had found a fractional charge and set out to explain the result using quantum mechanical equations, the rules that describe the motion and interactions of subatomic particles. A year later, he came up with a new theory, involving groups of electrons acting as a unit. He called these units quasiparticles. Individual electrons can couple to more than one quasiparticle, dividing their charge between the quasiparticles and giving the appearance that the electron has a fraction of a whole charge. In 1997 two groups of physicists saw direct evidence of quasiparticles, supporting Laughlin's explanation.

The effect that Stormer and Tsui discovered is known as the fractional quantum Hall effect. It does not have any immediate practical applications but may allow physicists to understand how electrons behaved at the beginning of the universe. It could play a role in the development of computer memories by using the fractional charges of electrons while they are parts of quasiparticles.

Gerardus 't Hooft

*"Gerardus 't Hooft Dutch physicist and Nobel prize winner. 't Hooft shared the 1999 Nobel Prize in physics with Martinus Veltman, for elucidating the quantum structure of electroweak, interactions in physics."*

# GERARDUS'T HOOFT

**'t Hooft, Gerardus** (1946- ), Dutch physicist and Nobel Prize winner. 't Hooft shared the 1999 Nobel Prize in physics with Martinus Veltman, also from The Netherlands. The two physicists developed mathematical calculations for predicting the structure and motion of subatomic particles. In honouring them, the Royal Swedish Academy of Sciences, which awards the prizes, stated that their work "has given researchers a well-functioning theoretical machinery which can be used for, among other things, predicting the properties of new particles."

Born in Den Helder, Netherlands, 't Hooft completed high school in The Hague in 1964. He studied physics and mathematics at the University of Utrecht, Netherlands, completing his undergraduate degree in 1966 and his Ph.D. degree in theoretical physics in 1972. From 1972 until 1974 he was a fellow at the European Laboratory for Particle Physics (CERN) in Geneva, Switzerland. He returned to the University of Utrecht in 1976 as an assistant professor and became full professor of physics in 1977.

The work for which 't Hooft and Veltman won the Nobel Prize began in 1969. As a 22-year-old student, 't Hooft studied high-energy physics with Veltman, who was then a professor at the University of Utrecht. 't Hooft and Veltman were looking for a theory that would link the four fundamental forces in nature, namely gravitation, electromagnetism, the strong nuclear force, and the weak nuclear force. The researchers' result, called the non-abelian gauge theory of electroweak interaction, linked two of the fundamental forces and gave the concepts of particle physics a firmer mathematical foundation.

The theory 't Hooft and Veltman created explains how elementary particles interact with one another, primarily how they interact through the electromagnetic and weak forces. The electromagnetic force governs how particles with electric charge interact, and the weak force governs how particles radioactively decay, or transform into different particles. Together they constitute the electroweak force. Experimenters have used machines called particle accelerators to confirm many of the theory's predictions, thus verifying its interpretation of electroweak interactions.

The Dutch scientists' calculations were critical in determining the mass of an elementary particle called the top quark, which was observed for the first time in 1995 at the Fermi National Accelerator Laboratory (Fermilab) near Batavia, Illinois. Their theory also predicts the existence of a particle called the Higgs boson, which should interact with other elementary particles to give the particles their mass. However, scientists have not yet detected the Higgs boson in an experiment.

Martinus J. G. Veltman

*"Martinus J.G. Veltman, Dutch physicist and Nobel Prize winner. Veltman shared the 1999 Nobel Prize in physics with his former graduate student Gerardus 't Hooft for elucidating the quantum structure of eletroweak interactions in physics."*

# MARTINUS J.G. VELTMAN

**Veltman, Martinus J.G.** (1931- ), Dutch physicist and Nobel Prize winner. Veltman shared the 1999 Nobel Prize in physics with his former graduate student Gerardus 't Hooft, who is also from The Netherlands. The two physicists developed mathematical calculations for predicting the structure and motion of subatomic particles. In honouring them, the Royal Swedish Academy of Sciences, which awards the prizes, stated that their work "has given researchers a well-functioning theoretical machinery which can be used for, among other things, predicting the properties of new particles."

Born in The Netherlands, Veltman received his Ph.D. degree in theoretical physics from the University of Utrecht, Netherlands, in 1963. He was a fellow at the European Laboratory for Particle Physics (CERN) in Geneva, Switzerland, until 1966, when he became professor of physics at the University of Utrecht. In 1981 Veltman was named John D. and Catherine T. MacArthur Professor of Physics at the University of Michigan. After retiring from the university in 1997, he settled in Bilthoven, Netherlands.

In 1963 Veltman developed the first general-purpose computer program for solving algebra problems. He designed the program to assist in the enormous calculations that are part of theoretical physics. Later algebraic programs have incorporated the ideas and principles of Veltman's program.

In 1969 Veltman and his graduate student 't Hooft began the work for which they would win the Nobel Prize. They were looking for a unifying theory that would link the four fundamental forces in nature: gravitation, electromagnetism, the strong nuclear force, and the weak nuclear force. The physicists' result, called the non-abelian gauge theory of electroweak interaction, linked two of the fundamental forces and gave the concepts of particle physics a firmer mathematical foundation.

The theory Veltman and 't Hooft created explains how elementary particles interact with one another, primarily how they interact through the electromagnetic and weak forces. The electromagnetic force determines how particles with electric charge interact, while the weak force governs how particles *radioactively decay*, or change into other particles. Together they constitute the electroweak force. Experimenters have used machines called particle accelerators to confirm many of the theory's predictions, thus verifying its interpretation of electroweak interactions.

The Dutch scientists' calculations were critical in determining the mass of an elementary particle called the top quark, which was observed for the first time in 1995 at the Fermi National Accelerator Laboratory (Fermilab) near Batavia, Illinois. Their theory also predicts the existence of a particle called the Higgs boson, which should interact with other elementary particles to give the particles their mass. Scientists have not yet detected the Higgs boson in an experiment. After retirement, Veltman continued his involvement in research to observe the Higgs particle.

Herbert Kroemer

*"Herbert Kroemer won Nobel Prize in physics in 2000 for developing semiconductor hetero-structure used in high-speed and opto-electronics."*

# HERBERT KROEMER

**Herbert Kroemer** (1928–), won Nobel Prize in Physics in 2000 with Zhores I. Alferov and Jack S. Kilby for basic work on information and technology in general and for developing semiconductor heterostuctures used in high speed and opto-electronics in particular.

Professor Kroemer received a Ph.D in theoretical Physics in 1952 from the University of Gottingen, Germany, with a dissertation on hot-electron effects in the then-new transistor setting the stage for a career in research on the physics and technology of semiconductors and semiconductor devices. Following work in a number of research laboratories in Germany and the USA, Kroemer persuaded the ECE Department at UCSB in 1976 to put the limited resources it had available for expanding their small semiconductor research program, not into main stream silicon technology, but in to the emerging compound semiconductor technology. In this field, Kroemer saw an opportunity for UCSB to become one of the leading institutions. He himself became the first member of the group, thus founding what has grown into a large group that is second to none in the physics and technology of compound semiconductors and devices based on them.

In his research, Prof. Kroemer has always preferred to work on problems that are one or two generation ahead of established mainstream technology. In the mid-50s, he was the first to points out that great performance advantages could be gained in various semiconductor devices (initially bipolar transistors) by incorporating what is now called heterostructures into the devices. Most Notably, in 1963 he proposed the concept of the double hetrostructure laser, the central concept in the field of semiconductor lasers, without which that field would simply not exist. These ideas were far ahead of their time and required the development of modern epitaxial growth technology, before they could become mainstream technologies, inturn providing a great stimulus towards the development of these technologies. Kroemer's receiving the Physics Nobel Prize in 2000 ultimately can be traced to these early papers. Only by 1980 the technology had progressed to the points that the 80's became a decade of "Heterostructures for everything a topic that continues to dominate compound semiconductors not just lasers and light-emitting diodes, but integrated circuits as well and is even invading mainstream silicon integrated circuit technology.

His current research is dominated by two quite different topics : Superconductor -Semiconductor hybrid structures involving InAs- AlSb quantum wells contracted by superconducting niobium electrodes. In such structures, the superconduting electrodes in essence induce super conductivity in the semiconductor of all semiconductors, InAs is the best materials for the construction of such devices.

Kroemer received several Prizes and Awards these are - J.J. Ebers Awards of the Electron Devices Group of the IEEE (1973); Heinrich Welker Medal of the International Symposium on GaAs and related compounds (1982); National Lecture, IEEE Electron Devices Society (1983); UCSB faculty Research Lecture (1985); Honorary Doctorate in Engineering, Technical University of Aachen, Germany (1985); Jack Morton Award of IEEE (1986); Donald W. Whittier Chair in Electrical Engineering (1986); Alexander von Humboldt Research Awards (1994); National Academy of Engineering (1997); Honorary Doctorate in Technology, University of Lund, Sweden (1998); Nobel Prize in Physics (2000).

Jack S. Kilby

*"Jack S. Kilby won Nobel Prize in physics for his part in the invention of the integrated circuit."*

# JACK S. KILBY

**Jack S. Kilby** (1923–), won 2000 Nobel Prize in Physics with Zhores I. Alferov and Herbert Kroemer, for basic work on information and communication technology.

There are few living men whose insights and professional accomplishments have changed the world. Jack Kilby is one of these men. His invention of the monolithic integrated circuit - the microchip - some 40 years ago at Texas Instruments (TI) laid the conceptual and technical foundation for the entire field of modern microelectronics. It was this break through that made possible the sophisticated high speed computers and large capacity semiconductor memories of today's information age.

Mr. Kilby grew up in Great Bend, Kansas. With B.S. and M.S. degrees in electrical engineering from the Universities of Illinois and Wisconsin respectively, he began his career in 1947 with the Centralab Division of Globe Union Inc. In Milwaukee, developing ceramic base, silk - screen circuits for consumer electronic products.

In 1958, he joined TI in Dallas. During the summer of that year working with borrowed and improvised equipment, he conceived and built the first electronic circuit in which all of the components, both active and passive, were fabricated in a single piece of semiconductor material half the size of a paper clip. The successful laboratory demonstration of that first simple microchip on September 12, 1958, made history.

Jack kilby went on to pioneer military industrial and commercial applications of microchip technology. He headed teams that built both the first military system and the first computer incorporating integrated circuits. He later co-invented both the hand-held calculator and the thermal printer that was used in portable data terminals.

In 1970, he took a leave of absence from TI to work as an independent inventor. He explored among other subjects, the use of silicon technology for generating electrical power from sunlight. From 1978 to 1984, he held the position of distinguished Professor of Electrical Engineering at Texas A and M University.

Jack Kilby is the recipient of two of the nation's most prestigious honors in Science and Engineering. In 1970, in a White House Ceremony, he received the National Medal of Science. In 1982, he was inducted into National Inventors Hall of fame, taking his place alongside Henry Ford, Thomas Edison, and the Wright Brothers in the annals of American Innovation.

Mr. Kilby holds over 60 U.S. Patents. He is a fellow of the Institute of Electrical and Electronics Engineers (IEEEE) and a member of National Academy of Engineering (NAE). He has been awarded the Franklin Institute's Stuart Ballantine Medal, the NAE's Vladimir Zworykin Award, the American Society of Mechanical Engineer's Holley Medal, the IEEE's medal of Honor, the Charles Stark Draper Prize administered by the NAE, the Cledo Brunetti Award, and the David Sarnoff Award. On the 30th anniversary of the invention of the integrated circuit, the Governor of Texas dedicated an official Texas historical marker near the site of the TI laboratory where Mr. Kilby did his work. In 2000, Jack Kilby was awarded Nobel Prize in Physics for his part in the invention of the integrated circuits

Eric A. Cornell

*"Eric A. Cornell won the Nobel Prize in physics in 2001, for the achievement of Bose-Einstein condensation in dilute gases of alkali atoms and for early fundamental studies of the properties of the condensates."*

# ERIC A. CORNELL

**Eric A. Cornell** (1961—), won the Nobel Prize in Physics in 2001 with Wolfgang Ketterle and Carl E. Wieman, for the achievement of Bose-Einstein condensation in dilute gases of alkali atoms and for early fundamental studies of the properties of the condensates.

A laser beam differs from an ordinary light bulb in several ways. In the laser the light particles all have the same energy and oscillate together. To cause matter also to behave in this controlled way has long been a challenge for researchers. This year's Nobel Laureates have succeeded - they have caused atoms to "sign in unison" - thus discovering a new state of matters the Bose - Einstein condensate (BEC).

In 1924 the Indian physicist Bose made important theoretical calculations regarding light particles. He sent his results to Einstein who extended the theory to a certain type of atom. Einstein predicted that if a gas of such atoms were cooled to a very low temperature all the atoms would suddenly gather in the lowest possible energy state. The process is similar to when drops of liquid form from a gas, hence the term condensation. 71 years were to pass before 2001 Nobel Laureates, in 1995 succeeded in achieving this extreme state of matter. Cornell and Wiemen then produced a pure condensate of about 2000 rubidium atom at 20nk i.e. 0.00000002 degrees above absolute zero.

Eric A. Cornell have B.S degree, Physics, with honor and with distinction, Stanford university 1985. and done Ph.D., Physics from MIT in 1990.

He is a fellow of JILA, NIST and university of Colorado at Boulder, 1994 to present. He is Senior Scientist, National Institute of Standards and Technology, Boulder, 1992 - to present. Also Professor Adjoint, Physics Department, University of Colardo, Boulder 1995-to present.

Eric A. Cornell received several Honors and Awards. Some of these are

- Member, National Academy of Sciences, 2000.
- Fellow, optical society of America, Elected 2000, R.W. Wood Prize, Optical Society of America, 1999.
- Lorentz medal, Royal Netherlands Academy of Arts and Sciences, 1998.
- Fellow, The American Physical Society, Elected 1997.
- I.I. Rabi Prize in Atomic, Molecular and Optical Physics, American Physical Society, 1997.
- King Faisal International Prize in Science, 1997.
- National Science Foundation Alan T. Waterman Award, 1997.
- Carl Zeiss Award, Ernst Abbe Fund, 1996.
- Fritz London Prize in Low temperature Physics, 1996.
- Department of Commerce Gold Medal, 1996.
- Presidential Early Career Award in Science and Engineering, 1996.
- New comb - Cleveland Prize, American Association for the Advancement of Science, 1995-1996.
- Samuel Wesley Startton Award, National Institute of Science and Technology, 1995.
- Firestone Award for Excellence in Undergraduate Research, 1985.
- National Science Foundation Graduate Fellowship, 1985-1988.

Wolfgang Ketterle

*"Wolfgang Ketterle won the Nobel Prize in physics in 2001, for the achievement of Bose-Einstein condensation in dilute gases of alkali atoms, and for early fundamental studies of the properties of the condensates."*

# WOLFGANG KETTERLE

**Wolfgang Ketterle** (1957–), won the Nobel Prize in Physics in 2001 with Eric A. Cornell and Carl E. Wieman, for the achievement of Bose-Einstein condensation in dilute gases of alkali atoms and for early fundamental studies of the properties of the condensates.

Dr Ketterle, Professor, now with Department of Physics Massachusetts Institute of Technology, received his diploma (equivalent to master's degree) from the Technical University of Munich in 1982, and his Ph.D. from the University of Munich in 1986. After Post doctoral work in molecular spectroscopy at the Max-Plank Institute for Quantum Optics in Garching (with Prof. H. Walther) and in combusion diagnostics in Hiedelberg (with Prof. J. Wolfrum) he went to MIT in 1990 as a research associate (with Prof. D.E. Pritchard). In 1993 he joined the physics faculty at MIT, and is now a full professor.

Wofgang Ketterle's research is in atomic physics and laser spectroscopy, particularly in the area of laser cooling and trapping of neutral atoms with the goal of exploring new aspects of ultracold atomic matter. Since the discovery of gaseous Bose-Einstein condensation, large samples of ultracold atoms at nanokelvin temperature are available. His research group uses such samples for various direction of research. Bose-Einstein condensates are a new quantum fluid. The interactions among the atoms make them an intriguing novel many-body system. Aspects of the interest are sound, superfluidity, and properties of miscible and immiscible multicomponent condensates. These topics are interdisciplinary with condensed matter physics.

The coherence properties of the condensate are exploited in the field of atom optics. Coherent beams of atom extracted from the condensate (atom lasers) are analoguos to optical laser beams. Ketterle's research group has used Bose Einstein condensates as amplifiers for light and for atoms. A third direction are precision measurements.

The unprecedented control over the position and velocity of atoms provided by Bose-Einstein condensates is exploited for high precision atom interferometry

Wolfgang Ketterle worked independentaly of Cornell and Wieman, and four month after them he reported large condensates of laser-cooled sodium atoms. Ketterle was able to demonstrate that the condensate actually behaved as a single coherent wave. He did an experiment where two stones are thrown simultaneously at a calm water surface and there wave patterns roll into each other, strengthening and weakening each other in systematic way. This is in greater contrast to what happens when uncoordinated matter, for example two fistfuls of sand are thrown on the water surface. Ketterle was also able to extract a beam of coherent matter from the condensate, thus achieving the first atom laser. An ordinary laser yields coherent radiations, an atom laser a stream of coherent matter.

When a gas consisting of uncoordinated atoms turns in to a Bose-Einstine condensate, it is like when the various instruments of an orchestra with their different tones and timbres, after warming up individually, all join in the same tone.

His awards include a David and Lucile Packard Fellowship (1996), the Rabi Prize of the American Physical Society (1997), the Gustav - Hertz Prize of the German physical society (1997), the Discover Magazine Award for Technological Innovation (1998), the Fritz London Prize in Low Temperature Physics (1999), the Dannie - Heineman Prize of the Academy of Sciences, Gottingen, Germany (1999), the Benjamin Franklin Medal in Physics (2000), and Nobel Prize in Physics in 2001, together with E.A. Cornell and C.E. Wieman.

Carl E. Wieman

*"Carl E. Wieman won the Nobel Prize in physics in 2001, for the achievement of Bose-Einstein condensation in dilute gases of alkali atoms, and for early fundamental studies of the properties of the condensate."*

# CARL E. WIEMAN

**Carl E. Wieman**, was born in March 26, 1951, Corvallis, Oregon. Carl E. Wieman won 2001 Nobel Prize in Physics with E.A. Cornell and Wolfgang Ketterle, for the achievement of Bose - Einstein condensation in dilute gases of alkali atoms, and for early fundamental studies of the properties of the condensates.

He received B.S. Degree from Massachusetts Institute of Technology in 1973, Ph.D. from Starford University in 1997 and Doctorate of Science (Honorary) from University of Chicago, 1997.

He was the Assistant Research Scientist, Department of Physics, University of Michigan 1977-1979, ; Assistant Professor of Physics, University of Michigan, 1979-1984; Associate Professor of Physics, University of Colorado, 1984-1987; Fellow, Joint Institute for laboratory Astrophysics (JILA) 1985 - Present; Professor of Physics, University of Colorado, 1987 - Present; Chairman JILA, 1993-1995; Distinguished Professor, University of Calorado, 1997 - Present.

He received many Honors and Awards, some of these are - Hertz Foundation Fellow, 1973-1977; Sloan Research Fellowship, 1984; University of Colorado Faculty Fellowship, 1990-1991; Guggenheim Fellowship, 1990-1991; Fellow of American Physical Society, 1990; Loeb Lecturer (Harvard University) 1990-1991; E.O. Lawrence Award in Physics 1993; Davisson - German Prize of the American Physical Society, 1994; National Academy of Sciences, elected 1995; Einstein Medal for Laser Science, Society for optical and Quantum Electronics 1995; Award for science, Bonfils - Stannton Foundation 1997; Lorentz Medal (Royal

Netherlands Academy of Arts and Science), 1998; Schawlow Prize for Laser Science (American Physical Society) - 1999; Phi Beta Kappa visiting Scholar 1999-2000; Benjamin Franklin Medal in Physics( Franklin Institute) 2000; NSF Director's Award for Distinguished Teaching Scholors, 2001.

Raymond Davis Jr.

*"Raymond Davis Jr. won the Noble Prize in physics in 2002, for pioneering contributions to astrophysics, in particular for the detection of cosmic neutrinos."*

# RAYMOND DAVIS JR.

**Raymond Davis Jr.**, was born in 1914, in Washington, DC, USA (US citizen). He received his Ph.D degree in chemistry in 1942 from Yale University, Connecticut, USA. He is a professor Emeritus at the Department of physics and Astronomy, University of Pennsylvania, Philadelphia, USA.

The Royal Swadish Academy of sciences has decided to award the Nobel Prize in Physics for 2002 with one half jointly to Raymond Davis Jr, and Masatoshi Koshiba, for pioneering contributions to astrophysics, in particular for the detection of cosmic neutrinos and other half to Riccardo Giacconi.

The Earth lies in the path of a continuous flux of cosmic particles and other types of radiation. This year's Nobel Laureate in Physics have used these very smallest components of the universe to increase our understanding of the very largest : the sun, star, galaxies and Supernova. The new knowledge has changed the way we look upon the universe.

The mysterious particle called a neutrino was predicted as early as 1930 by Wolfgang Pauli (Nobel Prize in 1945), but it would take 25 years to prove its existence (by Frederick Reines, Nobel Prize in 1995). This is because neutrinos, which are formed in the fusion process in the sun and other stars when hydrogen is converted into helium, hardly interect at all with matter and are therefore very difficult to detect. For example thousands of billions of neutrinos pass through us every second without noticing them. Raymond Davis Jr. constructed a completely new detector, a gigantic tank filled with 6000 tones of fluid, which was placed in a mine.

Over a period of 30 years he succeeded in capturing a total of 2000 neutrinos from the sun and was thus able to prove that fusion provided the energy from the sun. With another gigantic detector, called Kamiokande, a group of researchers led by Masatoshi Koshiba was able to confirm Davis results. They were also able, on 23 February 1987, to detect neutrinos from a distant supernova explosion. They captured twelve of the total of $10^{16}$ neutrinos that passed through the detector. The work of Davis and Koshiba has led to unexpected discoveries and a new intensive field of research, Neutrino Astronomy.

Masatoshi Koshiba

*"Masatoshi Koshiba won the Nobel Prize in physics in 2002, for pioneering contributions to astrophysics in particular for the detection of cosmic neutrinos."*

# MASATOSHI KOSHIBA

**Masatoshi Koshiba**, was born in 1926, in Toyohashi, Aichi, Japan (Japanese Citizen). He received his Ph. D. in 1955 at the university of Rochester, New York, USA. He is a Professor Emeritus at the International Center For Elementary Particle Physics, University of Tokyo, Japan.

The Royal Swadish Academy of sciences has decided to award the Nobel Prize in Physics for 2002 with one half jointly to Raymond Davis Jr, and Masatoshi Koshiba, for pioneering contributions to astrophysics, in particular for the detection of cosmic neutrinos and other half to Riccardo Giacconi.

The Earth lies in the path of a continuous flux of cosmic particles and other types of radiation. This year's Nobel Laureate in Physics have used these very smallest components of the universe to increase our understanding of the very largest : the sun, star, galaxies and Supernova. The new knowledge has changed the way we look upon the universe.

The mysterious particle called a neutrino was predicted as early as 1930 by Wolfgang Pauli (Nobel Prize in 1945), but it would take 25 years to prove its existence (by Frederick Reines, Nobel Prize in 1995). This is because neutrinos, which are formed in the fusion process in the sun and other stars when hydrogen is converted into helium, hardly interect at all with matter and are therefore very difficult to detect. For example thousands of billions of neutrinos pass through us every second without noticing them. Raymond Davis Jr. constructed a completely new detector, a gigantic tank filled with 6000 tones of fluid, which was placed in a mine.

Over a period of 30 years he succeeded in capturing a total of 2000 neutrinos from the sun and was thus able to prove that fusion provided the energy from the sun. With another gigantic detector, called Kamiokande, a group of researchers led by Masatoshi Koshiba was able to confirm Davis results. They were also able, on 23 February 1987, to detect neutrinos from a distant supernova explosion. They captured twelve of the total of $10^{16}$ neutrinos that passed through the detector. The work of Davis and Koshiba has led to unexpected discoveries and a new intensive field of research, Neutrino Astronomy.

Riccardo Giacconi

*"Riccardo Giacconi won Noble Prize in physics in 2002, for pioneering contribution to astrophysics, which have led to the discovery of comic X-ray sources."*

## RICCARDO GIACCONI

**Riccardo Giacconi**, was born in 1931, in Genoa, Italy (US Citizen). He received his Ph.D. in 1954 at the University of Milan, President of Associated Universities, Inc., Washington, DC, USA.

The Royal Swedish Academy of Sciences has decided to award the Nobel Prize in Physics for 2002 with one half jointly to Rayond Davis Jr. and Masatoshi Koshiba and the other half to Riccardo Giacconi for pioneering contribution to astrophysics, which have led to the discovery of Cosmic X-ray sources.

The Earth lies in the path of a continuous flux of cosmic particles and other types of radiation. This year's Nobel Laureate in Physics have used these very smallest components of the universe to increase our understanding of the very largest : the sun, star, galaxies and Supernova. The new knowledge has changed the way we look upon the universe.

The Sun and all stars emit electromagnetic radiation at different wavelengths, both visible and invisible light e.g. X-rays. In order to investigate cosmic X-ray radiation, which is absorbed in Earth's atmosphere, it is necessary to place instruments in space. Riccardo Giacconi has constructed such instruments. He detected for the first-time a source of X-rays outside our solar system and he was the first to prove that the universe contains background radiation of X-ray light. He also detected sources of X-rays that most astronomers now consider to contain black holes. Giacconi constructed the first X-ray telescopes, which has provided us with completely new and sharp images of the universe. His contributions laid the foundation of X-ray astronomy.